On the Sea Battle Tomorrow That May Not Happen

Dia-Logos

Schriften zu Philosophie und Sozialwissenschaften
Studies in Philosophy and Social Sciences

Herausgegeben von / Edited by
Tadeusz Buksiński / Piotr W. Juchacz

Bd. / vol. 25

PETER LANG

Tomasz Jarmużek

On the Sea Battle Tomorrow That May Not Happen

A Logical and Philosophical Analysis of the Master Argument

PETER LANG

Bibliographic Information published by the Deutsche Nationalbibliothek
The Deutsche Nationalbibliothek lists this publication in the Deutsche Nationalbibliografie; detailed bibliographic data is available in the internet at http://dnb.d-nb.de.

Library of Congress Cataloging-in-Publication Data
A CIP catalog record for this book has been applied for at the Library of Congress.

The Publication is funded by the Ministry of Science and Higher Education of the Republic of Poland as a part of the National Program for the Development of the Humanities 2016–2018, financed under grant No. 3aH 15 0186 83.

Project number: 0186/NPRH4/H3a/83/2016

NATIONAL PROGRAMME
FOR THE DEVELOPMENT OF HUMANITIES

Translated by Sławomir Jaskolski, Monika Smereka

Technical editors: Mateusz Klonowski, Krzysztof Krawczyk

ISSN 1619-005X
ISBN 978-3-631-74589-2 (Print)
ISBN 978-3-631-76079-6 (E-PDF)
ISBN 978-3-631-76080-2 (EPUB)
ISBN 978-3-631-76081-9 (MOBI)
DOI 10.3726/b14343

Peter Lang – Berlin · Berlin · Bruxelles · New York · Oxford · Warszawa · Wien

This publication has been peer reviewed.

www.peterlang.com

Książkę tę dedykuję moim Kochanym Rodzicom.
(*I dedicate this book to my Beloved Parents.*)
Tomasz Jarmużek

Foreword

This work is devoted to one of the most famous reasonings of the Greek logician and philosopher Diodorus Cronus. This reasoning was named *Master Argument*, probably due to the power it had on the audience.

It seems that the objective of the Master Argument was to substantiate a thesis that there are no pure possibilities within the empirical world, and hence, anything that is possible must happen sometime. This, in turn, suggests some form of determinism: since there are no such possibilities that are not, nor ever would be, realised.

Diodorus' argument had not survived to our times. It is known solely from the sources that describe its subject, premisses and thesis quite laconically.

There are many reasons, not just historical ones, to believe that, by means of his argumentation, Diodorus disputes with Aristotle. For in his classical reasonings about The Sea Battle Tomorrow, the Stagyrite defended indeterminism by allowing for the pure possibilities, ergo that it is possible for the sea battle to happen as well as it is possible that it shall not happen.

For these reasons, the attempt to reconstruct and analyse Diodorus' argument should refer to the reasoning of Aristotle, since both theoreticians have different views on the same matter — the nature of possibility, future and determinism, at the same time remaining within the same ancient logical–philosophical paradigm of reasoning.

With this observation in mind, I have decided to approach the reconstruction of Diodorus' argument with respect to what is already known about *the tomorrow's sea battle* of Aristotle. Therefore we shall discuss Diodorus' argument but in the context of the ancient dispute on determinism. Hence the title of this book: *On Sea Battle Tomorrow that may not happen. Logical and philosophical analysis of Master Argument*

The book, however, is not only of a historical nature. The history of ancient arguments, often relatively unknown or known from the comments of subsequent doxographists, constitutes merely a background for more modern reasoning. While adopting the findings of analytical philosophy and the tools of formal logic, these arguments may even diverge from historical facts, but they can develop the substantive elements highlighted by the historical sources.

Apart from the analyses carried out, a certain accomplishment of the study is a demonstration that, in the light of modern logic, Diodorus' argument can be reconstructed without assuming determinism, that is, without accepting the linear time structure. Thus, as we claim, The Sea Battle Tomorrow may not happen, as well as it, of course, may happen. So, luckily, not all things have been decided yet.

Acknowledgments

This book constitutes a further improved and translated version of the Polish book originally titled: *Jutrzejsza bitwa morska. Rozumowanie Diodora Kronosa*, published in 2013 by The Nicolaus Copernicus University Press in Toruń. Its present translation is financed under grant No. 3aH 15 0186 83, English translation and edition of book *Jutrzejsza bitwa morska. 3aH 15 0186 83 Rozumowanie Diodora Kronosa*,' awarded by the Minister of Science and Higher Education of the Republic of Poland. Hereupon, I would like to render thanks for the financial support and the decision to grant it.

As the author of the book, I would also like to express my great and hearty thanks to the people who, to various extents, scientifically contributed to the preparation of this text. In the alphabetical order, they are: prof. dr hab. Elżbieta Kałuszyńska; prof. dr hab. Jacek Malinowski; dr hab. Marek Nasieniewski; dr hab. Zbigniew Nerczuk; dr Maciej Nowicki; prof. dr hab. Andrzej Pietruszczak; and also not named colleagues from the Department of Logic, NCU.

Additionally, I would like to thank the doctoral students under my tutorial: my associates, Mateusz Klonowski and Krzysztof Krawczyk for they have undertaken the thankless and Sisyphean tasks of the technical editors, and due to their assistance, I have avoided many errors. I am very excited by the fact that these young researchers, as well as the book, remain in the tradition of the Polish School of Logic. For me, this is of a superior value.

I would also like to acknowledge the work done by the translators of the book: Mr Sławomir Jaskólski and Mrs Monika Smereka, who did their best to convey the essence of the book without breaching the rules of English grammar.

And finally the last, but at the same time one of the most important recipients of thanks, is my Family.

In the first instance, I am very grateful to my Beloved Parents: Elżbieta and Mirosław Jarmużek. The reason is simple and logical (sic!): if it wasn't for Them, I would not exist, and so would this book. But another important reason for this vote of thanks is more complex. My Parents always loved me and took care of me to such an extent which I will never manage to return. Therefore, this book is dedicated to Them.

Of course, I am also grateful to my Beloved Wife Joasia who — in spite of her own very important challenges — devoted a lot of time for me, so I could finish this book. And since I mention Joasia, I cannot forget our children: Helena and Kazimierz. Whilst acknowledging my family, I would also like to mention my brother Piotr and his wife Asia. Their assistance allowed me to save plenty of time and energy for the work on the book. I really appreciate that.

There is, of course, a great deal of people and institutions I should express gratitude to, but then, the very content of the book would not stop to mount up. Therefore, I shall take a certain ontological shortcut. I will, thus, thank God. Without Him, there would not be even these words of thanks in any shape.

I owe a debt of gratitude to the listed persons. At the same time it is clear — certainly — that any possible faults are solely attributable to the author.

Contents

Part I Philosophical framework of the topic

Part I Philosophical framework of the topic

1 Introduction

Imagine the following scene. Here, a man of a reverend appearance, immersed in a philosophical reflection, stands by a picturesque, rocky gulf of ancient Greece. That gulf had already been a theatre of many sea skirmishes in which the Athenians wrestled with their enemies. Its image, therefore, naturally brings up associations with the former battle scenes. This is when Aristotle — who is the reverend man — affected by that scenery and cultivated in contemporary Greece *bios theoretikós*, asks the famous question: *will there be a sea battle tomorrow?*

It is quite possible that this scene occurred. What is certain, however — if we resist a philosophical temptation to raise immoderate objections and stick to the common sense plane[1] — is that Aristotle actually uttered this famous sentence in such circumstances or he did it elsewhere. All in all, we believe that things of the past belong to a field which may not be affected by anyone or anything. Once shaped, it cannot be altered.

The philosopher's question did not, however, pertain to the past, but to the future. This question was to illustrate the issue of whether or not the expressions stating something about the future events hold any logical value and thus whether, while making statements about future, we can *a priori* reasonably believe them being are true or false.

Aristotle was to negatively resolve the problem he faced, opting for the open, underspecified future. He displayed his position in Chapter 9 of *On Interpretation*, supporting it with the reasoning which then found numerous attempts of interpretation and reconstruction with the use of modern, logical measures. The purpose of this reasoning — as it seems — was to clearly distinguish the field of what used to be from the field of what may or may not happen in the future (further in this book, the argument of *On Interpretation* shall be often referred to as **RA**, the abbreviated *Reasoning of Aristotle*).

Also Diodorus Cronus, a Megarian logician and philosopher, took up this issue in ancient discussions, probably trying to justify a position that the events that are to take place in the future are preordained in a way[2]. To this

1 cf. subsection 1.3.7. *Truth and time vs. epistemic concepts.*

2 It was argued here that, within his arguments with Aristotle, Diodorus attempted to positively resolve the problem of the sea battle (Prior A. N. [231], s. 138). However, it can be believed that he only meant a temporal characteristics of modal expressions

end, Diodorus made use of a special reasoning (further in the study referred to as **RDC**, abbreviated *Reasoning of Diodorus Cronus*)[3].

Future[4] features — among other things — this difference from the present that, in common experience, it simply *does not exist*. The more so, there are no events, states of the world, affairs or facts etc.[5] that will or may be in the future[6]. They may indeed — in the best case — only occur.

If however — as Diodorus could have argued — the expressions referring to future events (i.e. events which only will occur!) bear a logical value, the events described by them must be in some way — proportionately to the logical values of the expressions — preordained, even before they occur. Preordained and therefore inevitable and determined.

It would therefore appear that from the objective perspective, the reasoning of Diodorus is in favour of certain concepts of the time and the world that, while

---

which in ancient times were perceived as leading to a version of determinism. In principle, both subsequent parts of the study are devoted to these problems.

3 The historical, philosophical and logical contexts of this discussion as well as the preserved passages of Aristotle considerations and Diodorus' reasoning, together with the necessary interpretations shall be exhaustively presented in the next part of the study.

4 When using words: *past, present, future*, I do not opt for the position of realism in the philosophy of time, corresponding to the so-called $A-series$ ordering according to McTaggart, and thus I decline the position which corresponds to the $B-series$ ordering. I only use these words as natural language constructs by which we globally define what used to be, what is and what will be. The issue of McTaggart's complex problem and details related to *A and B series* with reference to the question of Diodorus reasoning shall be further outlined in Chapter 3.

5 For identification of the objective language, I shall use the following expressions: *state of affairs event* or *occurrence* (and accordingly, their verb forms: *occur* and *happen*). I am going, however, to operate them very neutrally — just like for the words: *past, present* and *future* — without opting for any of the ontological schools, but only considering what we are capable of describing using constatives of a natural language. Moreover, I shall discuss this issue in more detail in Chapter 1 where I am going to more evidently advocate for one of the positions.

6 When saying *do not exist*, I only mean that they do not exist at present. If they did, they would have already taken place by now, and not in the future, should the circumstances allow. This remark is obviously very general and preliminary. Many events — or maybe even all the events, depending on the approach to the structure of the world — are tensely *extended*. In this very sense, due to the existence in present, they also *extend* into the future. But before they take place in the future, they may only exist in the present at most, as some stirrings of what is going to occur at a later time.

ensuring a logical value of an expression on the future, lead to a certain version of the fatalism.

Let us briefly look into the issue of expressions. Regardless of whether or not their logical value is preordained before the events described do (or do not) take place, they must contain a time parameter. **RDC** should therefore take into account both the time of given expression and the section or point in time described thereby. From the linguistic perspective, a reflection on **RDC** theory must therefore contain certain logic which takes into account the relationship between given language units and the time, more specifically — its appropriate concepts.

On the other hand, in the case of Aristotle's concept, language and its relationships with the time and the world, expressed in certain explicit and implicit assumptions, are not supposed to determine the logical values of any sentences, but basically leave underspecified at least some of the sentences that describe the future. **RA** must therefore advocate for a certain asymmetry between the descriptions of the past, present and future world, as they lead to a version of anti-fatalism, hence to a world open to variations of the future events.

In the case of fatalism, the logical values of appropriate language units must of course be based on their relevance against the world described (and in our case, the relevance against the future world). The correctness of expressions gain particular importance as the theory of truth in a classic bivalent approach is also a theory of falsity.

These above preliminary findings show that a study on the concept of Diodorus must give a primary role to the notion of *truth* and the notion of *time*.

In this case, however, these notions are involved in various relations. Diodorus tried to show that:

1. expressions to describe the future are already true or false in the present time, which means that:
2. the future is closed in a way, determined.

It seems however that for the expressions about the future to be true:

3. the future events must be specified, determined.

The third notion that inevitably appears is therefore the notion of *determinism*. And *a contrario*, while advocating for logical indeterminacy of statements describing the future, Aristotle was an advocate of open future and thus of the *indeterminism* of the future.

These very four notions taken together form the overall conceptual framework within which the philosophical discussion between Aristotle and Diodorus is located.

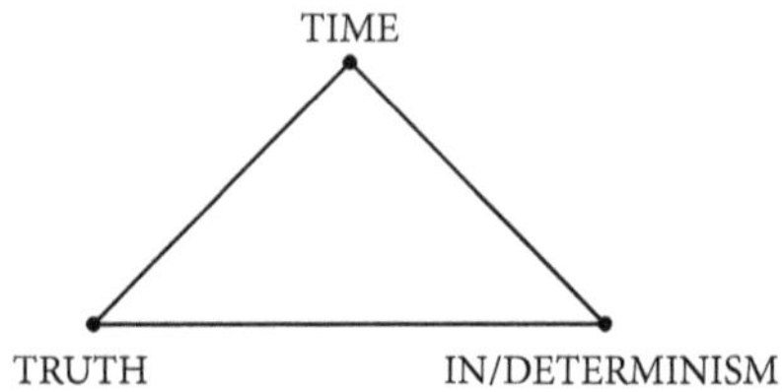

The above diagram does not give precedence to either of the listed concepts. Only the adoption of the relevant theoretical perspective honours some of them, making them more primary. In view of the chosen perspective, the study gives a particular importance to the concept of *time* characterised in a specific way, probably familiar to the ancients in a way. However in the case of **RDC**, it must remain in the defined relationship with the other two concepts. Therefore, the first part of the study shall discuss these three basic components and outline them for two reasons:

1. due to the need to situate the ancient discussion in a broader context which will enable explanation of its role and purpose in a rather complete manner, and outlining a number of additional problems that we encounter when describing the philosophical context;
2. because of the need to clarify the theoretical assumptions of the study — assumptions without which it would be infeasible to precisely take up the key topic and its further analysis.

When commenting on the individual components, I will try to isolate them as far as possible in order to maintain the general character of the reflection. Of course, this will be neither entirely possible nor desirable. Since the concepts that mark out the framework for the problems of Diodorus reasoning are interconnected, discussing each one separately requires necessarily references to the others. During the analysis and the description of each of them, I will try to determine the expected connection with the subsequent part of the reflection.

2 Truth and Sentences

In the present chapter, we shall introduce the issues associated with the language-world reference. Needless to say, the majority of the issues are of a general nature and are the topics of philosophy of language, semiotics and linguistics. However, some appear directly as a result of the considerations relating to the subject of this study, thereby increasing the spectrum of the general concepts. This reason alone is sufficient to justify the fact that many of the concepts will be omitted, some barely mentioned, and certain ones — only indicated. However, the whole chapter constitutes a compact attempt to outline the problematics and the notion of language in terms of the notion of time; it identifies the arising problems, and highlights and presents the assumptions further considered indisputable.

The language we use on a daily basis has many functions. It can be used to express feelings, create new situations, and change others' attitudes and beliefs. However, carrying out an individual function requires fulfilling a more primary one, commonly referred to as the descriptive–informative function.

Being a part of a human community and benefiting from it requires the participants to be involved in the communication process. This process involves passing information encoded in the language about the broadly defined world. The language is thus a medium enabling communication. Efficient communication depends on various factors. One of them is a matter of understanding a statement or decoding information.

However, communication is not a goal per se. Of particular note is thus — as previously mentioned — the information to be transferred. Statements are particularly valuable when conveying a lot of information on the subject of communication. However, the statements may be carrying little information or even be inconsistent with reality, thus misinforming.

In order to increase the understanding of given statements and to reduce their ambiguity, artificial languages are being created or natural languages clarified. These languages are structured in such a way as to best serve the aim of describing particular objective domain.

Among many classifications dividing statements in particular language, one of the most interesting philosophical (and not just) aspects is the division into true and untrue expressions (false or devoid of logical value).

For the purposes of the study, I would like to determine a few arrangements of the concept of statements, their meanings (content transmitted), logical values and finally, I would like to refer these issues to the concept of time. In the subsequent sections of the study, these arrangements shall allow reference to certain issues without their additional clarification.

2.1 Syntactic approach

The deliberations on statement should begin with defining the broad term sometimes used within this chapter. The term in question is *declarative statement*. By *statement*, we mean: *any graphical sequence of phrases in a particular language, or certain inscription or its phonetic equivalent; compliant with the grammatical structure of given language*[7].

According to the above definition, the statement does not have to be a sentence in a grammatical sense. However, for quite obvious reasons (that will also be clearly named in this chapter), the statements of interest are either sentences or certain abridgements of sentences, as in some context their utterances fulfil the same informative function as certain sentences. Such statements will be hereafter named *declarative statements*[8]. Obviously, by the above definition, statements are related to a particular language used to articulate them, for grammatical and syntactic rules form a language. For example, the following inscriptions:

(1) *It is raining.*
(2) *Haben Sie etwas zu machen?*

are statements — respectively — in English and German. As mentioned, sometimes certain phrases not meeting the criteria of syntactic category of a sentence in their informative function are equivalent to other certain statements that belong to the sentence category, namely:

(3) *a fifth grade student*

in its informative function corresponds to the sentence:

(4) *A fifth grade student won mathematics competition.*

---

7 Thereby, we exclude those elements of communication that do not comply with the grammatical rules. I assume, however, that if these carry any meaning, it can be articulated with the use of some statement. Obviously, this means that if something cannot be articulated with the use of some statement, it does not carry any meaning.

8 As you can see, presented approach is not purely syntactic. For syntactic approach considers sentences (and other expressions) of particular language as finite sequences of simple phrases (cf. e.g. Tokarz M. [281], pp. 11–13), without referring to their informative function, therefore also to their content.

if it aims to answer the question:

(5) *Who won mathematics competition?*

In order to determine this, it is necessary to step outside the purely syntactic aspect[9] and consider given phrase in pragmatic aspect. This issue will be discussed later.

From the syntactic point of view, the sentences can be divided into simple, compound sentences and complex sentences. This division is based on the quality and quantity of sentence-forming functors, present in a given sentence.

The definition of the simple sentence or compound sentence can therefore read as follows. By *simple sentence*, we usually mean such a sentence which *comprises exactly one sentence-forming functor; functor of clauses that are not sentences*[10]. On the other hand, *a complex sentence* is a sentence *consisting of at least one sentence-forming functor; clause's functor*.

In the course of further studies on **RDC** and its logical reconstructions, the subjects of the analysis will be sentential expressions, which, in terms of information, correspond to both simple and complex sentences. Two more following comments shall be added to the issues of syntactic nature.

Firstly, although the term sentence can be used interchangeably with the term sentential expression, only the term sentence will be used in further parts of the study. Such measures are justified by the fact that almost any given sentential expression corresponds to an equivalent informative sentence. The issue of how it can be done will be introduced and discussed in due course. Each use of the term sentence in other meaning, especially broadening its content, will be adequately highlighted.

Secondly, as we focus on the sentential expressions only in their informative function, only constative sentences will be considered. Imperative sentences,

---

9 Actually, the same applies to other types of statements. For example, rhetorical questions are used in common language in order to express beliefs, rather than ask a question. Sometimes even individual words, such as *fire*, uttered in right context, can be equivalent to other statements; in this case to: *There's a fire*! This example comes from Ajdukiewicz K. ([7], p. 27).

10 This definition is explicitly based on the concept of categorial analysis of Kazimierz Ajdukiewicz. It will though not be reminded, as it constitutes permanent and active elements of philosophical and logical culture. Therefore, I will not be introducing the definitions of functors — either functor-forming, name-forming or sentence-forming, as these terms can be found in the works of Ajdukiewicz ([3, 4, 8]).

interrogative sentences and others will be omitted, even though these may also comprise informative components. However, if they contain informative component, we assume that can be expressed in constative sentences.

2.2 Meanings of expressions

Sentential expressions are not only correctly structured sequences of simple phrases. Due to their informative function, many have some — more or less — specified content[11]. That is because the phrases, that compose these expressions, usually also have specified meanings. Let us see the following sentence:

(6) *It is snowing outside.*

This sentence analysed as an expression, in the common meaning of its individual simple phrases, transmits the information that *it is snowing outside*. However, certain sentences seem to be — despite correct structure — meaningless, like the sentence:

(7) *The human mind weighs one kilogram, twelve grams.*

For in its common meaning the word *mind* refers to a certain and abstract object, not associated with the concept of weight and unrelated to it, or at most excluding it.

However, the term *mind* could be considered as having the meaning of the word *brain*. According to this interpretation, the sentence (7) would convey information that *human brain weighs one kilogram, twelve grams*. No doubt, this sentence would thus make sense, although, seen as a general sentence — false.

This example shows that the meaningfulness of a sentence depends on the meaning assigned to the individual simple phrases.

2.2.1 Propositions as meanings of sentences

Our reflection on the issue of meaning will start with the reflection of Willard Van Orman Quine who wrote that so-called semantic aspects (not only in linguistic meaning, but also in the logical one) divide into two groups of concepts, leading to certain ambiguity in the use of the word *semantics* ([233], p. 178). The first use entails issues relating to the concept of the meaning of the phrases and can be an alternative to the term *the theory of meaning*. I will currently focus on these

11 Or meaning, or sense.

particular issues, while the other use of the term *semantics*, which meaning is synonymous to *theory of reference*, will be discussed in the next subsection.

Philosophy and logic is used to discuss the meaning of the simple phrases or corresponding notions. In accordance to the sentences structured from them, propositions correlated to those phrases are mentioned. Therefore, propositions are the combination of meanings or concepts corresponding to simple phrases[12]. The idea of meaning has been though developed in philosophy in various ways and with diverse consequences for such understood proposition.

The theories of meaning, not from just historical but also substantial point of view, have approached the problem of meaning from various angles. The division based on the problem of existence of meanings and propositions seems to be essential. Most of all, it is associated with the question how the meanings exist, thus what kind of objects they are from ontological point of view? The answers for this question have varied.

On the grounds of psychological approach (also known as associationism), meaning has been comprehended as a certain mental idea associated with a given word. This approach originated mostly from modern British empiricism, following the idea of John Locke [161]. An attempt to reduce the meaning to the specific mental experiences has resulted from nominalistic approach. However, the key issue of associationism is a question if certain phrases correspond to single mental processes (of particular person and in particular moment) or to some types of mental processes, not dependent on a given user of the language. Since mental experiences of various people during communication process vary, meanings, defined as mental experiences, must therefore be subjective and relativised to a certain situation, and any use of the phrase would have been giving it a new meaning. On the other hand, discussion about certain types of mental processes corresponding to a given phrase, results in losing its nominalistic nature. It places though the meaning in a realm dependent on the language users, but not related to any particular person. As psychologism advocated for

12 Hereinafter, we will see that propositions expressed by sentences are not always functions of the meaning of the simple phrases used to structure the sentence. This issue has been emphasised by Frege in words: *never to ask for the meaning of a word in isolation, but only in the context of a proposition* ([75], p. 10). When it comes to relation between proposition and ideas, there has been a dispute resulting in isolating two viewpoints on the subject: ideas priority (known as allogenic theory) and antithetic, stating that propositions are objects *sui generis* (known as idiogenic theory) (cf. Woleński [295], pp. 33–34). However, we are more interested in determinations not related to the origin of propositions, but their function in transmitting information.

naturalism in explication, it causes an issue, as discussing the types, moves the problem to the field of idealism.

The other approach to the issue of meaning is based on the connotation concept, initiated by John Mill [192]. From its point of view, the meaning of phrases has been looked for from the perspective of physical objects, by means of the concept of connotation, comprehended as a set of monosemous characteristics of objects corresponding to the range of the appellation. The meaning (of general names, but not proper names) has been understood precisely as connotation.

Apart from non-linguistic area, either mental or physical, some concepts also view the meaning as an abstract and ideal object. This approach places the meanings in the similar area as mathematical objects within Platonism in philosophy of mathematics. The contents of the names, or propositions expressed by sentences are to carry the same ontological status as distributive sets, numbers, etc. This approach originated from the philosophy of Plato, modern times and present days, developing in philosophy of Alexius Meinong [187], Edmund Husserl [112], and *sensu stricto* logically, in writings of Gottlob Frege [76, 77] and Alonzo Church [52].

On the grounds of his phenomenological philosophy, Husserl had developed a concept of intentionality of meaning, that is an act of directing the subject to derogate the meaning of a sign, that is object of meaning. This concept has been studied by Roman Ingarden, who qualified the meanings as certain abstract objects, called intensional objects. However, in Ingarden concept, these objects differ from Husserl's ideal objects, e.g. by having their own beginning in time, whilst ideal objects are not characterised by any time coordinates. [115].

The greatest impact on the development of logical semantics came from works of Frege. The main ideas of his works *Über Sinn und Bedeutung*, on the meaning can be presented as follows. *Sinn* is a sense of an expression, whilst *Bedeutung* is a referent of an expression, ergo what it refers to. According to Frege, name phrases express their senses which forms a synonym of the term *notion*, in other words, correspond to certain ideas (*Der Begriff*). However, sense does not guarantee existence of the recognised object (therefore the reference of the appellation does not have to exist), but most importantly, it differs from individually associated projections of phrases. On the other hand, *proposition in logical meaning* (*Das Gedanke*) is a sense of a declarative phrase, it also differs from associated projections of sentences. All true sentences designate logical values called true, and all false sentences designate logical values called false. Sentences also constitute a certain types of names. Direct speech sentences (*oratio recta*) are appellations of logical values and indirect speech sentences (*oratio obliqua*), prefixed with e.g. intensional functors[13], are the appellations of certain propositions. The following statement illustrates this difference:

(8) *John was lying, when he said, he saw Stan.*

With a certain understanding of the word *lie*, it does not matter if the sentence *John saw Stan* is true. Logical value of dependent clause is thus irrelevant. What is important is that the phrase *lie that p* means:

i) to say that p,

and

ii) to be sure that the case is not like p claims.

In the above case, dependent clause is not a name for logical value, but for a certain proposition.

Frege's concept expresses objectivistic approach to proposition, as the meaning expressed by sentence. This approach is constantly present in modern logic and philosophy, in spite of many various nominalistic and subjectivistic alternatives. In my study, while considering examples of various sentences, when it comes to both meaning of simple phrases and propositions, I will be inclined to objectivist approach[14].

Obviously, the theory of meaning not only was but has constantly been developed in philosophy, particularly in analytic philosophy. In addition to that mentioned above, many other conceptions of meaning have been proposed, e.g.: as intensional structure (Carnap R. [46].), as use of phrases (Wittgenstein L. [292]), by directives of acceptance of sentences (Ajdukiewicz K. [5]), by knowing verification conditions (Carnap R. [47]). Suffice to say that already in the work *The Meaning of Meaning* published in 1923, Ogden and Richards [206] offered about sixteen meanings of the word *meaning*, although some are associated with cultural approach to semiotics.

The problem of the theory of the meaning of linguistic phrases involves not only the question of denotation of a term *meaning* but also the extent of what can be a meaningful phrase. Individual appellations constitute separate issue — as long as they are not abbreviations of certain descriptions —

---

13 Intensional functors create complex sentences, so their logical value is not unequivocally determined by logical value of simple sentences.

14 Naturally, it does not mean that I believe that objective sense of a sentence is and must be always directly grasped.

they name specific and unitary objects in the universe of discourse, without conveying any meaning[15]. This issue is related to a more general problem of proper names, or phrases performing functions of naming or indicating an individual in order to singularise it. Such names may include occasional designations, such as: *this*, *that*, *these* etc. According to certain philosophers, proper names should though preserve the autonomy of assigning objects to conditions e.g. by Tadeusz Czeżowski [54]). Basically, it seems difficult to draw a demarcation line between proper names and other phrases.

Therefore the conclusions from the discussion are neither unambiguous nor generally accepted. One of the reasons for the above is surely the fact that natural language constitutes a complicated and multifaceted formation in constant use. Hence, new linguistic phenomena, resulting from e.g. new ways of use of certain archaic phrases, constantly emerge.

For the purposes of the study, devoted directly to the issue of logical phrases expressing the states of objects, that are to take place in various time intervals or moments, both past and future; the above considerations are fully reasonable. The classical approach to the concept of ontology of truth, which will be discussed in due course, postulates that by having logical value, a statement expresses certain information about specific objective conditions, which are or not, accordingly. Therefore, determining logical value of a statement depends in the first instance on identifying expressed proposition, thus determining the state of the objects according to the statement.

Surely, at the moment I am not going to outline the general theory of meaning[16], but simply employ certain existing ideas. Partly historical above considerations enable forming some binding reflections in the subject.

The categories of meaning and proposition can be considered on various grounds, most of all in relation to purely psychological area or to abstract objects domain.

---

15 However, some philosophers believed that individual names had meanings. For example, Duns Scotus used a term *haecceitas* as a certain elementary essence characterising individuals. Others, in turn, claimed they had no meanings (Mill J. S. [192], Russell B. [247], Dąmbska I. [56]).

16 It would be difficult, especially on the grounds of modern discussion, where in the context of, amongst others, Quine's criticism of meaning and his concept of indeterminacy of translation, the attempt to define meanings of sentences as propositions; or propositions as meanings of sentences, may be classified as a vicious circle, and surely is not satisfying (Grayling A. C. [86], pp. 27–28). Current state of our presentation allows to clarify general concepts, so to be able to use them without unnecessary confusion in solving other problems.

In the first case, these should be considered as appropriate mental processes present in perceptorium of specific users of particular language *hic et nunc*, thus essentially as unrepeatable conscious mental acts (Ajdukiewicz K. [5], pp. 109–120) or as types of such processes. Within this approach, meaning and proposition should be considered in a psychological sense.

On the other hand, these ideas can be also seen as not dependent upon individual mental experiences. This approach enables autonomizing the information conveyed within a phrase from individual subjective mental processes. Thus understood meaning of phrases and sentences is logical. The meaning attached to a constative sentence in a language is therefore usually considered in this approach as proposition in logical sense. (Ajdukiewicz K. [2], pp. 148–149).

In my study, I accept the objective view in some generality. For we focus on the sentences expressing information on the world, and these are in essence independent on persons who utter them. From a purely theoretical point of view, the information may not even constitute a content of any particular sentential expression nor many literally different ones.

Propositions are expressed by sentences, or — more generally — by sentential expressions. Therefore, the term *sentence* will be used not only in a syntactic sense, but in a broader way: *sentence* will mean *a simple or a complex sentence of a language, that, in certain understanding, can be assigned a certain proposition or particular idea relating to the world in the means of reporting*[17]. This definition should reflect the intuitive meaning of broadly accepted fact that constative sentences say something about the world, and it depicts some part of the world.

Therefore, when using the term *sentence*, I will mean not only a certain syntactic unit of given language, but also a correlated proposition that refers reportedly to some state of the object[18].

2.2.2 Pragmatic components of a statement

Identification of propositions from a given sentential expression is rarely simple and direct. It is not only due to the issues with polysemy of individual phrases

17 This definition has been formed based on the definition of constative sentence by Ajdukiewicz K. ([7], p. 28).

18 I do not call this kind of proposition a logical proposition. For it does not have to be either true or false, and these terms usually are reserved for propositions carrying logical value. Many sentences about future, discussed below, doubtlessly express specified propositions, but — considering Aristotle's concerns presented in the second part of the book — do not have to carry any logical value.

and consequently — sentences but it also involves also the sensitivity of majority of statements to the context of their utterance.

The context of a statement utterance may affect uttered phrases in such a way that literally the same sentences or phrases will induce different propositions[19].

This subject is being developed nowadays as part of logical pragmatism. Published in 1938, work of Charles Morris *Foundations of the theory of signs* [196] has had a significant influence on the formation of this discipline. In his study, Morris divided semiotic research into three subdisciplines: syntactics, semantics and pragmatics. Pragmatics was to analyse relations between language and its users (phrases, marks, statements, etc). In the objective field of pragmatics, among others, the following issues are discussed: casualness phenomenon, sentential attitudes, presuppositions, conversational implicature and speech act (Tokarz M. [282], p. 108).

In pragmatic research, sentence as certain abstract grammatical unit is distinguished from its utterance by particular person in certain time and circumstances. For in terms of meaning, the meanings of sentences in a natural language depend on a context. Thus, meanings conveyed through them do not have to be a function of meaning of simple phrases. Only knowing a context allows to identify proposition expressed and to establish their logical value. Let us consider the following statement:

(9) *He is lying to you.*

Sentence (9) does not express any proposition unambiguously. For it is not clear who stands for the personal pronouns used in the sentence (9). Therefore their referents are unknown. Finally, the time of the lie is also unknown.

---

19 This is also — in a diachronic aspect — due to a linguistic practice and its historical and social conditioning, that change the meaning of the phrases and introducing new meanings. In a classic approach, propositions are designed as eternal objects insensitive to changes: they do not come into being nor disappear. (cf. Lycan W. G. [164], pp. 76–87). Disappear or come into being can at most languages or phrases used to express the propositions. For this study, it is not necessary to opt that far in the characteristic of propositions. Claiming that they differ from sentences and sentential expressions would suffice. The fact that the latter change their meanings due to conditions suggests a conclusion that they express different propositions at that moment (if still carry any sense at all). In this case, when analysing the sentence, it is handy to assign them appropriate indexes, which correspond to different propositions expressed. The above-mentioned approach to propositions is therefore instrumental. The classic approach will be further narrowed in the last section of this chapter, where we will discuss the epistemic concepts of truth.

Whether it is a current period of time (at this moment? this year?), past tense, future, or some continuity (e.g. lying for a month?). Without the context of (9), it is not possible to answer any of the above questions, nor identify the proposition expressed in the statement. (9) seen as abstract does not express any proposition; therefore, it cannot be neither true nor false.

To each such sentential expression can be attached a list of questions aiming to supply so much missing information to determine the proposition expressed. There is no way to create a general list of questions allowing to determine the contexts of all the statements. However, creating a complete list of relevant questions for each particular sentential expression could be possible. Establishing the context provides information required to form the sentence equivalent to the sentential expression, that is not a sentence. For example, for the expression:

(10) *Fire!*

information needed could come from the following list of questions: *What is on fire?*, *Where the fire takes place?*, *When the fire takes place?* Knowing the answers to these questions allows us to formulate an equivalent informative sentence:

(11) *(Object x) (in place y) is on fire (in time of z).*

by inserting phrases corresponding to the transmitted information in brackets[20]. By supplying missing information, a statement is being created; it is not sensitive to the aspects of statement context which have been expressed explicitly. For example, we could get a sentence:

(12) *Private office of the Prime Minister in Warsaw is on fire on 20 May 2010.*

The context should of course provide information on the meaning of *to lie* or *fire*, if the phrases do not convey a clear meaning. If so, to the above list there should be also added a question: *What does 'lying' mean?* or *'fire'*. This problem applies particularly to the indistinct terms and ultimately a good criterion to decide if a given object qualifies to a given meaning. However, I shall not focus on this problem, assuming that if any declarative statement expresses a proposition, thus shows reporting attitude towards certain state of affairs, then the criteria to establish the statement's logical value can be explicated by operationalising it in a certain way.

20 Such information can be called *values of parameters*, and an action of supplying *parameterisation.*

To sum up, knowledge of the context provides information that enables determination of the proposition expressed in a given statement. The proposition can be then expressed with a statement insensitive to a context. Logical value of the statement determines the knowledge of the world.

This study will be based on the viewpoint that propositions should meet at least the first of two following conditions described by Gottlob Frege. Frege postulated that proposition should have permanent objective reference ([77], p. 52). An issue of objective perspective of the language is currently omitted. Let us look at this postulate from a linguistic perspective. The same grammatically understood sentence may be used in different meanings or contexts, expressing different propositions. For example, in the expression:

(13) *Stan is a frog.*

we can say that certain animal called Stan is a specimen of frogs, or teasingly and a bit poetically state that Stan (a specimen of *homo sapiens*) is a good breaststroke swimmer.

Therefore, in the above examples, this sentence is the same from grammatical point of view, but differs from the expressed content's, thus propositional, point of view. This is effective in different objective respect, as in each case the sentence is about completely different objects and characteristics.

In everyday practice, the majority of statements are of the same nature as the above examples. Therefore, while analysing statement pragmatic components, complementing a given statement with necessary information should be considered.

Within logical pragmatism, one of the most important studies is *Indexical expressions* by Yehoshua Bar-Hillel [21], where the subject of logical pragmatism has been situated within natural language as an investigation of relations between linguistic phrases and their specific use in determined contexts. Further studies on the structural point of view of the context can be found in the works of Richard Montague: *Pragmatics* [193], *Pragmatics and intensional logics* [194] and in the article of Dana Scott *Advice on modal logic* [257] (this approach will be discussed in due course). These issues have been discussed also by David Kaplan in *On the logic of demonstratives* [130] and by Max J. Cresswell in *Logics and Languages* [50].

Moreover, Scott and Montague in their works proved a redundancy of logical propositions in an interpretation of colloquial sentences — sentences do not express propositions, but their specific utterances. In opposition to this opinion and pro necessity to include propositions in his work *Pragmatics* argued Robert Stalnaker [263]. The development of modern pragmatics issues and research on the context have substantially benefited from works of Saul Kripke [138, 139, 140], on relational semantics of possible worlds, used to clarify the definition of context and interpretation of intensional phrases[21].

The problem of context is directly connected with the second condition laid down by Frege, that is that the logical value of logical proposition should remain fixed[22] ([77], p. 52). For in various contexts, the same proposition could carry seemingly different logical value. For example, sentence:

(14) *I am in Toruń.*

seems to change the logical value depending on the context of its utterance. For in different contexts, it expresses different propositions. According to Frege's idea, logical value of propositions does not change under any circumastances. Therefore, occasional and all other expressions of which the list of missing information is not completed, will express different propositions, depending on their parametrisation. According to Frege's beliefs, if those propositions differ in logical value, then they are not identical, but carry different information on characterised domain.

Note that the principle established above is only an implication. It means that it is possible that a certain statement in different circumstances carries the same logical value, while expressing completely different proposition. Instead of sentence (14), sentence 'I am here' could be investigated, as it is true in every context, but almost every time, after parametrisation, it expresses different proposition. In the study, we shall accept a possibility that the same proposition carries a different logical value in different time contexts. The acceptance of such possibility shall be reasoned in due course.

Obviously, the problem of identifying the expressed proposition is purely pragmatic, not theoretical. I assume that in many cases, it is possible

21 Transformation procedures from certain statement to its logical value, through context, proposition and possible world shall not be described within this study. It can be found e.g. in work of Jacek Malinowski *The Pragmatic interpretation of utterances* [175]. In this article, the author considers the problem of arranging, under logical pragmatism, two definitions of meaning: as proposition expressed in a statement or as logical value denoted by a statement.

22 As it clearly arises, other important issue on logical value is a question about what carries logical value — statements or propositions. Obviously, Frege opted for propositions. Nowadays, many philosophers believe otherwise. I will address this issue and take a position on it in one of the next subchapters, meanwhile tentatively using terms *logical value, true, false*, either relating to sentences expressing propositions, or to propositions itself.

to translate phrases, dependent semantically upon the context, into sentences independent of it, or aka eternal sentences. In the works related to pragmatics, on occasional phrases, there is some criticism towards belief that for each expression dependent on the context there is an eternal sentence, equivalent in meaning; considering casual phrases as natural elements of colloquial language (cf. Bar-Hillel Y. [21], Levinson S. [158]). However, I will not try to draw a demarcation line between occasional phrases that can be replaced by certain eternal sentences, and ones that cannot. Certainly, for many statements of constative character, after parametrisation, there can be given sentences expressing, in particular context, the same information as the statement. Moreover, this is the reason why these statements are called sentential expressions.

Let us get back to the issue of context formally developed by Dana Scott. In the above-mentioned article *Advice on modal logic*, the author has provided a definition of the context of uttering given sentence. The definition gives guidelines on factors affecting the information carried in the statement. According to Scott, the index, or formal representation of the context, is a certain coordinate n as follows:

$$\langle w, t, \langle x, y, z\rangle, a, \ldots\rangle.$$

where w means possible world, t moment of time, trio $\langle x, y, z\rangle$ Cartesian coordinates of a position in space, a a person forming a statement. Free space in coordinate n is for other important parameters of a given context.

As can be seen, in this definition, time is one of the parameters. In eternal sentences, the value of this parameter is usually given in the sentence and does not need to be considered to clarify when given state of affairs occurred. See examples below:

(15) *On the 15th of July 1410 Polish and Lithuanian forces won the Battle of Grunwald.*
(16) *It is raining in Toruń.*

The first example expresses proposition independent of time of the statement, in terms of content of communication (obviously not necessarily in terms of its logical value). The second example is dependent on time when (16) is uttered. Uttered on 13th of October 2002 means that *on 13th of October 2002 it is raining in Toruń*. More generally, at any time uttered t, means that *at the time t it is raining in Toruń*.

In studying the reasoning of Diodorus, both types of statements will be useful. Therefore, corresponding categories used in subsequent analyses shall be defined.

2.2.3 Sentences that are temporarily determined

While considering various phrases in terms of syntactics and semantics, we have formulated a definition of a sentence, based on idea of proposition. We have also indicated various pragmatic circumstances related to assigning propositions to certain phrases. However, I have assumed that in many cases, those circumstances may be taken into account in a suitable sentence independent of context, expressing the same proposition as the statement dependent on the context. We are mostly interested in the aspect of time expressed in a sentence; therefore, this type of sentences we would like to name as *sentences determined in time.*

As *sentences determined in time*, we will mean simply sentences expressing any state of affairs at determined in them time[23]. A set of such sentences will be assigned a symbol $\mathbf{S}_{TD}$[24].

2.2.4 Sentences that are temporarily undetermined

Another type of sentences represent such statements as (16). Parametrizing them gives us sentences determined in time. For further purposes of the study, we shall need a category of a sentence parametrised by all the aspects except time. Completing such sentence with a missing value of a time parameter gives us a sentence of $\mathbf{S}_{TD}$.

Procedurally, specifying the value of time parameter should be considered as assigning a variable, ranging over set of certain time objects of given value[25].

Sentences of above-mentioned type will be called *sentences undetermined in time*, their set marked with symbol $\mathbf{S}_{TU}$[26].

---

23 Determining time may be of various accurateness, e.g. sentence 'Tomorrow Jan will go to work.' does not specify when a given state of affairs will take place; but the sentence 'Jan will go to work after 20th of April 2003.' is under my definition of a sentence determined in time. Detailed consideration on different types of sentences and their reference to time will be provided in the further part of the book.

24 The concept of such sentences is similar to Quine's concept of eternal sentences, that is sentences which logical value is constant through contexts ([235], pp. 221–222).

25 Formal conceptual apparatus to represent the defined sentences and time shall be presented in further parts of the book, featuring a formal analysis of ancient reasoning. We will also reflect there on the domain for time variable. For it requires — as you can see — certain form of dating.

26 The concept of these sentences corresponds to the sentences analysed in the Ancient times and in the Middle Ages. These sentences can change their logical values in relation to different time frames. This particular concept of sentences has been used in creating logic of time by Arthur North Prior ([230], p. 8).

Two more remarks are worth making. Firstly, sentences belonging under $\mathbf{S_{TU}}$ do not express propositions, as they do not have permanent objective reference. Therefore, they are neither true nor false[27]. Secondly, given any $s_1 \in \mathbf{S_{TU}}$ and adding two different time variables $t_1 \neq t_2$, we receive two different sentences determined in time $\langle s_1, t_1 \rangle \neq \langle s_1, t_2 \rangle$ that express different propositions and may carry different logical values. Given any $s_1 \in \mathbf{S_{TU}}$ and adding any value of a time parameter, we receive such s_1' that $s_1' \in \mathbf{S_{TD}}$ and of course it expresses some proposition. Therefore, adding a value of time parameter can be seen as complementing sentences undetermined in time with time aspect of the context.

Assume that both defined sets are nonempty. Moreover, they sum up to a set marked as **S** that henceforth will be called simply a set of sentences. Its elements split in a disjoint and independent way into sentences determined in time and sentences undetermined in time.

2.3 Sentences *vs.* logical value

Insofar discussed the concept of meaning as a content expressed by particular sentence, or in general terms statement, is rather of pragmatic-linguistic nature. However, according to above-mentioned Quine's remark, semantics in a narrow sense, can be seen as denotational theory.

Accordingly, in logic there is a different concept of meaning than expressed by particular utterance of content or proposition. In terms of logic, meaning represents a logical value of given sentence, which it carries in its model. This concept derives from ideas of Gottlob Frege[28].

However, not interpreted sentences are not able to carry logical value. For they are not associated with any propositions and do not say anything about any subject. Although, they can carry different logical values under different interpretations. This problem is evident in the analysis of constructed language, such as a language of First Order Logic (hereinafter: **FOL**). Let us consider the following phrase **FOL**

(17) $\exists_x \, p(x, y)$

---

27 For they are of similar nature as open formulas in classical logic of quantifiers, i.e. a logical value depends on the interpretation of free variables.

28 Frege distinguishes between the above-mentioned *Sinn* and *Bedeutung*, or sense from denotation. True sentences indicate logical value of truth, and false sentences — of false [76].

If the formula (17) is interpreted in reference to a natural domain, and predicate p is attributed a relation of 'smaller than' $<$, then its logical value will depend solely on a denotation of free variable y. For denotation function assigning y number 0, formula (17) will be false because natural number smaller than 0 does not occur. On the other hand, for any denotation function that to y attributes natural number other than 0, the formula (17) will be true. However, if (17) is referred to the domain of real numbers in relation of 'smaller than', then for any denotation function, this formula will be true.

Also in natural language, the same phrases can — as mentioned above — carry various informative content. The following sentence:

(18) *The sun is sitting on the sofa.*

depending on the interpretation can be senseless or senseful: false or true. All depends on the proposition associated with a given sentence. Obviously, interpreting phrases from natural language is not as precise as interpreting formulas in a constructed language. However, when the proposition expressed in a given sentence is known, we can try to establish its logical value.

2.3.1 Requirements for the concept of truth

Stating that certain statements are true or false, we usually operate some concept of truth. This issue becomes more important when we notice that we will be working on sentences expressing propositions on the states of affairs that take place in various times, particularly in future. Therefore, we need to specify few matters of logical value that will allow analysis of various sentences without any restrictions imposed on the value of time parameter.

The remarks of Bertrand Russell [250], about the concept of truth, additionally amended, should prove useful. First of all, the concept of truth must be also the concept of false. In addition, it should specify the carrier of truth (and false), and the reason why the carriers of truth are true (or false).

2.3.2 Historical background

The issue of the concept of truth should be considered from the perspective close to historical one[29], in an attempt to indicate the constituents of the concept

29 This discussion does not aspire to be called comprehensive. According to the experts on the subject, the full outline of the issue of truth has not yet been studied (Woleński J. [295], p. 175).

of truth that seem key for the study, and clarify the close proximity where the outline of concept of truth in demand is located.

The term *truth*, in different stylistics is often being used in different language contexts. Suffice to say that Polish or English speakers speak[30] frequently about *true opinions*, *true things* (e.g. *real, true picture*), *true convictions*, *true beliefs* etc. However, sometimes they also speak about *true propositions* or *true sentences*. The aim then is the propositional understanding of truth carriers (Woleński J. [295], p. 172). This very technical sense of the word *true*, in reference to propositional objects constitutes the matter of philosophical analysis within discussion on the concepts of truth.

Basically, the established concepts of truth can be exhaustively divided into epistemic and non-epistemic (Künne W. [142], p. 145). The epistemic concepts lead the question of truth of sentences to rely on the cognising subject and conditions in which it can state if the sentence is true. On the other hand, the non-epistemic concepts are based on a belief that logical value does not depend on the cognising subjects, but entirely on the objective perspective, that is a necessary and sufficient condition for adequate sentence to carry a logical value. Based on the non-epistemic concepts, the knowledge of the subject about the logical value of a given sentence is irrelevant. For presumably, there are many sentences of never known logical value. For example, the sentence:

(19) *Napoleon Bonaparte one minute before his death had an odd number of hair on his head.*

Intuition and historical knowledge suggests that this sentence has a logical value: it is true or false. However, it is not possible to clearly identify its logical value, as we do not have in our command appropriate cognitive qualities and probably we never will. Perhaps the reason for this is also indistinctness of terms used in the sentence (19).

In the book, we will rather focus on the non-epistemic concept of truth. The epistemic concepts will be mentioned in due course as they often provide grounds for discussion on the concept of logical value of sentences about the past as well as the future.

The non-epistemic concepts of truth can be divided into relational and non-relational ones (Künne W. [142], p. 145). The essence of this division is based on the issue of explication of the idea of truth. The relational concepts, in *definiens* express belief that truth is to be a bi-argumental relation between

30 Obviously, much like other languages users.

the linguistic and objective perspectives, indicating a certain applicability, adequacy or consistency. On the other hand, the non-relational concepts do not use the term of bi-argumental relation *truth* or *false* in relation to the logical value carriers.

Among supporters and also precursors of the relational approach to the question of truth, first of all we should mention Saint Thomas Aquinas. This philosopher wrote that *(…) For, since the truth of the intellect is 'the adequation of intellect and thing,' inasmuch as the intellect says that what is is and what is not is not* ([278], I vol. ch. 59, pp. 127–128). The above-mentioned equation is usually described by formulary *adaequatio intellectus et rei*. Another work of Saint Thomas states: *(…) For this reason truth is defined by the conformity of intellect and thing; and hence to know this conformity is to know truth* ([279], quest. 16, art. 2, p. 249). The issue of agreement has been discussed also by Kant who claimed that the truth is *the agreement of cognition and its subject*, using the term cognition in a meaning of *proposition* ([126], s. 102).

On the other hand, philosophers formed also non-epistemic and non-relational concepts of truth[31] who without trying to define the truth based on the idea of relations between its carrier and objective perspective. One of the initial outlines of such a concept can be found in Plato's *Sophist*, where the Athenian philosopher states that a true sentence[32] states *something is, as is*, on the other hand, false *one states things that are other than the facts(…) In other words, it speaks of things that are not as if they were* ([222], 263B, p. 93)[33]. Interestingly, the Ancients are ascribed the use of such grammatical constructions that require some supplementation, or — as named in this case — parameterisation, to express proposition unambiguously. In particular, such sentences, according to given definition, can be elements of a set of sentences undetermined in time $\mathbf{S}_{\mathbf{TU}}$ (cf. Kü W. [142], pp. 146–155), e.g.:

(20) *Socrates leads a discussion.*

---

31 Detailed in this study concepts of Plato and Aristotle have been considered non-relational by e.g. the above-mentioned Künne W. [142], pp. 146–155.

32 Plato sees sentence as logical value carrier, mentioning relationship between noun and verb. However, the type of those sentences is not specified([222], 261C–262E, pp. 89–92). Similar understanding had Aristotle (Reale G. [236], pp. 533–536); however, only in his writings, more precise considerations on the types of sentences that can be true or false, can be found ([11], 17a, p. 71).

33 Translated literally from Greek it means: *states nothing as being* (translator note, p. 93).

In any case, this issue shall be characterised closer, while describing semiotic beliefs of Diodorus Cronus, as it will be of importance.

Plato's concept has been then developed by Stagyrite, giving a full weight and clearer conceptualisation to the classical definition of truth. The famous passus 1011b of *Metaphysics* reads as follows:

> To say that what is, is not or that what is not, is, is false; but to say that what is, is and what is not, is not, is true ([14], 1011b, p. 99).

The above passage can be understood i.a. as follows: statement that it is not as it is, or that it is as it is not, is false. On the other hand, a statement that it is as it is, or it is not as it is not, is true.

The Aristotle concept permanently joined European scientific thought, forming basis for further deliberations on the problem of truth. In the past century, Alfred Tarski essayed to polish it, consciously adverting to the idea of Aristotle[34]. In his works, he separated objective language from metalanguage, thereby indicating how to avoid semantic antinomies[35].

Alfred Tarski described a measure to define the predicate *true*. Because in his conception, the carriers of truth are to be constatives (not propositions), therefore *truthness* needs to be in reference to the language L. The definition of truth for language L should meet the general principle called T-convention, of the following form:

(T) *Sentence x is true in language L if and only if p,*

where x stands for metalanguage appellation of the sentence of objective language, and p is its translation into the metalanguage.

According to Tarski, T-convention is not in fact the definition of truth, but a methodological principle regulating the construction of the definition of truth in the language:

> We can only say that every equivalence of the form (T) obtained by replacing 'p' by a particular sentence, and 'X' by a name of this sentence, may be considered a partial definition of truth, which explains wherein the truth of this one individual sentence consists. The general definition has to be, in a certain sense, a logical conjunction of all these partial definitions ([272], p. 346).

---

34 Tarski adverted to i.a. previously cited passage from *Metaphysics*, writing that none of heretofore given conceptualisations *is sufficiently precise and clear* ([272], p. 233).

35 In particular with regard to the so-called liar paradox ascribable to Eubulides of Miletus (cf. Tarski A. [271], §1, [272], §7, §8), that seems to strike especially, according to the above-introduced classification, classical version of the concept of truth.

However, the method of constructing the definition of truth given by Tarski, is not appropriate for a colloquial language, as there are many levels of language, often undifferentiated (as colloquial language is a so-called language *semantically universal*), and its structure is not clearly defined, what can easily lead to semantic antinomies ([271], p. 14). According to Tarski, *the meaning of the problem* [i.e. defining truth for natural language] *is more or less vague*, and *its solution can have only an approximate character* ([272], p. 370). Tarski's concept of truth can be primarily used with constructed languages, as a theory of the conditions of correctness of expressions in a given language. The key quality of the discussed concept of truth is that it does not provide a criterion to determine if given sentence is true or not ([273], p. 317). It only provides methods to construct definition of truth for a given formalised language[36]

While outlining the concept of truth for the types of sentences we will be working on, we will consider guidance on levels of language, however, bearing in mind that we are primarily interested in sentences in colloquial language.

Addressing this issue, we will point to another problem. It arises from relativisation of the idea of truth to a given language and is related to the question on the carrier of logical value.

2.3.3 What owns the truth?

Amongst the possible candidates to the carriers of the truth or false — sentences understood obviously as constative sentences and expressing certain idea, content) and propositions, which are what the constative sentences express — were mentioned most frequently.

Following the famous argument of Alonzo Church [52] (cf. Grayling A. C. [86], pp. 20–21), we will lean towards the view that the carriers of logical value are propositions. Church argued that different sentences (e.g. from different languages) carry exactly the same logical values despite completely different forms of writing. What gives them identical logical value is the fact that they express the same proposition[37]. The logical value is determined by the

36 Or for the language meeting various formal conditions, such as: a full dictionary of expressions, strict syntactic rules relating to the structure of the sensible phrases in the language, unambiguity of functions performed by given language units, constant logical value for similar phrases independent of a context, or lack of casual phrases etc.([273], pp. 313–314.).

37 This argument is quite convincing, even for authors declaring reluctance to accepting the existence of such objects as propositions (cf. Woleński J. [295], p. 173).

expressed propositions[38]. However, it does not mean that it is not possible to mark sentences e.g. true. Some of modern authors write that truth is implicitly and imitatively also a property of sentences (cf. Barwise J., Etchemendy J. [24], p. 26), and the same can be probably said about false.

In the study, I will not be writing about propositions, but *truthness* or *falseness* will rather mark sentences, provided that according to the above-mentioned definitions, the sentences are expressing certain propositions and they co-determine logical value of the sentences.

They co-determine since the problem of logical value of sentences, according to one of the presented postulates on the theory of truth, can be considered from a non-linguistic point of view. For within non-epistemic concepts of truth, truthness or falseness of the sentences is mostly a consequence of objective aspect of discourse.

2.3.4 What makes sentences true

That certain sentences are true or not is decided by certain objects: It raises a question: what types of objects are they?

This reference to a given objective domain is the reason for proposed answers to the above questions to be included in so-called ontology of truth (Mulligan K., Simons P., Smith B. [199], p. 287).

Regardless of whether in given concept in explication of the word *true* or its different stylistic forms, requires a binary relation (between the logical value carriers and something that belongs to the objective aspect), or not[39], true sentences must state something, or mean something (e.g. indicate *a state of affairs*, cf. Tarski A. [272], p. 223).

In order to determine these objects, philosophers introduced different theoretical terms, usually referring to well-developed concepts describing — in the words of Russell — *the kind of thing that makes a proposition true or false* ([249], p. 185). Following Peter Simons when writing that: *classical theory of truth requires certain essence for simple statements about the world,* he adds: *well known under various names, such as 'status', 'condition', 'process' and 'occurrence' (…), the majority of them are the truthmakers* ([258], p. 159).

---

38 Therefore, propositions have been assigned two out of three traditionally assigned to them roles: meaning of sentence and carrier of logical value. Third one is the subject of propositional attitude (see Hart W. D., McGinn C. [98], p. 299).

39 What — as mentioned above — is to be a criterion of a division into relational and non-relational concepts.

Amongst many proposals for the so-called *truth-makers*, or *verifiers*; further I shall use *verifiers*), there are i.a. facts.

The concepts of the facts have been presented in various ways by: George Edward Moore [195] — as simple theory of facts and Bertrand Russell [249], who has developed the concept of *truth-makers* so it could correspond to different levels of sentential complexity, singularising unitary objects, different levels relations and the facts that are to correspond to complete sentences. Not without an impact on Russell's theoretical ideas was*The Tractatus Logico-Philosophicus* of Ludwig Wittgenstein, where the author introduced his so-called picture theory of language. Let us analyse the following theses from *The Tractatus*:

1 The world is all that is the case.
1.1 The world is the totality of facts, not of things.
2 What is the case — a fact — is the existence of states of affairs.
3 A logical picture of facts is a thought.
3.01 The totality of true thoughts is a picture of the world.
4.024 To understand a proposition means to know what is the case if it is true. (One can understand it, therefore, without knowing whether it is true.) (Wittgenstein L. [291]).

The concepts of *truth-makers* are therefore clearly metaphysical[40]. For discussing the description of the world from linguistic point of view, they introduce certain objects constituting the structure of the world.

Wittgenstein's beliefs have had a major impact on Russell's considerations, especially from the days of *The Philosophy of Logical Atomism*. However, in the later period — as verifiers *verifier* — instead of the facts Russell introduced other elements of the structure of the world, called *occurences* ([248], pp. 284–288).

Apart from the above-mentioned candidates for the truth-makers (verifiers), many other objects with their appropriate concepts have been proposed. Also as verifiers situations were proposed, the idea introduced by Roman Suszko[41] and developed by Bogusław Wolniewicz [296]. The situation was here understood as each fragment of *reality, verifying a sentence* (p. 14). The theory of situation has been also developed by Jon Barwise and John Perry [25].

As for newer proposals, we should mention so-called moments of which concept has been formulated by already-mentioned Mulligan, Simons and

40 On a side note, by *metaphysics* I mean the ontology of what exists, or real world. On the other hand, the ontology is a wider discipline, which domain is a set of possible objects, not only existing ones.

41 Drawing inspiration from Wittgenstein's Tractatus (cf. Omyła M. [208]).

Smith[199]. Obviously, the word *moment* has been understood by them in a completely different way than it will be in further parts of the book. By *moments*, they meant objects existentially independent but requiring existence of other objects. The above-mentioned authors give the following examples of moments: *smiling lips* or *a slip of a man* as examples of truth-makers for following sentences consecutively: 'A face smiles.', 'A man slips.' (p. 290). The concept of moments developed from works of Husserl (in any case, the term moment, as the authors admit, has been taken from Husserl) about the terms of ontological dependence and independence, and associated with them problem of wholeness and partiality (p. 294).

Apart from the signalised above proposals, there is one more concept that sparks philosophers' attention. By this, I mean the concept of state of affairs. On the basis of this concept, Ludwik Borkowski formulated the definition of truth as follows:

> The sentence is true if and only if it describes a certain state of affairs (claims existence of a certain state of affairs) and this state of affairs exists (happens) ([36], p. 119).

This definition imposes the two following conditions on the accuracy of given sentence: Firstly, the sentence must state something. It has to be a state of affairs. Secondly, the described state of affairs must exist. These guidelines, as will be seen further, will prove to be extremely valuable while we start to analyse sentences describing future or past states of affairs, i.e. the states of affairs that surely do not exist currently.

In the concept of truth, outlined below, for truth-makers the term *state of affairs* will be used. Although the term will not be here closely defined — as it would require broader discussion much beyond the subject of this book — however, it shall become clearer during case study. It can be assumed a primitive notion.

Finally, we will enunciate one more important remark. The verifiers, no matter how understood, do not make all the types of true sentences true. For there are such sentences of which the logical values seem independent of any objective conditions. Such sentences are usually called analytical (respectively true or false). Although there are many definitions of analytical sentence, quite often of different ranges (one of the most comprehensive lists of various approaches can be found in the work of B. Mates [177]), however, the cofeature of such sentences is that their logical value is based on two matters: structure and meaning of phrases present in it. Probably the most popular example of a sentence analytically true in English is as follows:

(21) *Every bachelor is an unmarried man.*

If we tried to parametrise such sentence, stating e.g. that:

(22) *In place x, in time t each bachelor is an unmarried man.*

then, irrespective of the values of x and t, the sentence would still be true (although of course expressing different proposition, less general). For sentence (21) states that *any bachelor is a man without a wife* and, according to classical[42] approach to analytical sentences, its logical value does not depend on any objective conditions. Therefore, introducing additional restrictions cannot have any influence on its logical value. That truthness (or respectively, falseness)

of analytical sentences does not depend on any state of affairs, is the reason why they will not be interesting, thus not taken into account.

However, for the issues discussed in this book, the sentences of logical value dependent on the objective perspective are important. Therefore, we will be interested in sentences usually called *synthetic*, or those that doubtlessly increase our knowledge on the empirical world, stating that *things in the world are so in some time*. These are e.g. sentences as follows:

(23) *Last Friday Mars extra-close opposition took place.*
(24) *On 12th of December 2007 Jan Kowalski won 2 million zloty in Lotto.*
(25) *Today I am totally overfed, after having two plates of zur kujawski prepared by my mum.*

2.3.5 The concept of truth employed in the study, concept of false controversy

We have established that propositions will be the carriers of logical value, and the states of affairs will be truth-makers of propositions and sentences expressing them. We will now proceed to attempting to create a certain outline of the definition of truth and to indicating the issues associated with the corresponding definition of false.

A substantiation for the use of truth-makers in the proposed definition of truth is the following attempt to apply to discussed question T-convention, as follows:

42 I use term *classical*, since it is known that nowadays some authors object to the division of sentences into analytical and non-analytical, as indistinct; and to the *analyticalness* of analytical sentences as based on non-analytical source in the shape of determining process, stabilising the semantic convention for synonymous phrases. I mean here particularly the criticism of Quine [234]. In my study though, I favour the approach I have described as classical, at least in the operational extend.

(26) *For any simple sentence* $z \in \mathbf{S}_{\mathrm{TD}} : f(z)$ *is true if and only if z.*[43]

Obviously, in this case, we are not interested in the relativisation of sentences with $\mathbf{S}_{\mathrm{TD}}$ to a language, in which the sentences have meaning — for I have assumed that to $\mathbf{S}_{\mathrm{TD}}$ sentences expressing proposition are assigned. Let us substitute *z* with the following questions:

(27) *Jan is going to school in moment t.*
(28) *Jan went to school in moment t.*

Let us assume that (27) is uttered in moment *t*, and (28) in any other later moment[44]. Then with the use of (26) we get:

(29) *'Jan is going to school in moment t.' is true if and only if Jan is going to school in moment t.*

and

(30) *'Jan went to school in moment t.' is true if and only if Jan went to school in moment t.*

This is a strange consequence, as one would like to acknowledge that the sentence (28) in some sense *results* from sentence (27), therefore the reason why the sentence (27) is true, makes also the sentence (28) true. Sentence (27) is true, when Jan goes to school in moment *t*, but what if in next moment, say t', Jan has already arrived at school. For we will not decide that the sentence ceased to be true. Its trueness is based on the fact that in moment *t* the state of affairs is that Jan is going to school; in t' the state of affairs may be different, but this is not important any more. Now it is clear why the sentence (28) *results* from (27). For they relate to the same state of affairs, if one is true, the other is true too, but in different time[45].

43 Where $f(z) =$ '*z*', i.e. f is a function assigning to $z \in \mathbf{S}_{\mathrm{TD}}$ its any metalanguage appellation '*z*'. This procedure is applied to prevent objections of some authors to quantifying over objects of type *x*, where *x* is not a non-dependent phrase, but only a graphic component of the phrase of type '*x*' cf. Woleński J. [295], p. 217, Black M. [33], p. 326). However, in contexts other than definitions of truth or false, even predicating a quantification, for simplification, I will use just '*x*'.

44 For the purposes of the presentation of the question of truth and false, I use quite intuitive term moment. In the chapter on the issues related to time and in the further chapters, I shall try to clarify them.

45 In the discussed concept, in the first place, the propositions are true, and imitatively, the sentences. Meanwhile, within semantic definition of Tarski, the types of inscriptions of given language are considered, or meaningful sentences, not their

Why then both sentences differ from grammatical point of view? The difference is of course in the grammatical tense forms. One is expressed in present tense, the other one in past tense. It arises from the position in time of the person uttering (27) and (28) in relation to the state of affairs expressed in the sentences. In the first example, the sentence is formulated in time t' approximately equal to t, and in the second example, in time t'' later than t. This observation gives some insight into the nature of tenses and functions they fulfil. This problem will be analysed in more detail in due course.

Apparently the definition of truth should not prefer any position in time. It means that a true sentence (if true) is simply true regardless of whether it states about the past, present or future states of affairs.

Let us look into the issue again, from the objective perspective. The states of affairs are organised in time. For example now, when I write these words using a computer, it is 20 : 45 Central European Summer Time, but just five minutes ago, when I ate delicious apple pie, called *staropolski jabłecznik*, it was 20 : 40 Central European Summer Time. The individual states of affairs I have encountered can be organised by the direction of time flow. The same applies to the future states of affairs I will be participating in. They can also be seen as organised in time considering binary relation of*being earlier* or its converse *being later*[46].

The sentences belonging under $\mathbf{S}_{\mathrm{TD}}$ express occurrence of certain state of affairs in time defined in them. Such understanding of sentences enables forming the following definition of truth, that seems *not to be sensitive* to time of utterance of the sentences:

(31) For any simple sentence $z \in \mathbf{S}_{\mathrm{TD}}$:$F(z)$ is true if and only if the state of affairs marked by sentence z happens.

Taking the sentence (27) in its different (in terms of grammatical tense) variant forms, we find that it is true if and only if in t the state of affairs is so Jan

---

meanings (or in our terminology propositions) (cf. Woleński J. [295], p. 228). The above problems result also from uncritical use of T-convention, without paying attention to the types of the logical value carriers and — as will be shown below — the ratio of time of statement to the transmitted content.

46 Note that this relation depends on the reference system of measurement. For as known, within modern physical time science, the time is not absolute. This issue will be discussed further, in the section devoted to *time*. However, for the purposes of the current deliberations, the time shall be considered as absolute — from certain reference system.

goes to school. The word *goes* means here present tense in relation to described state of affairs, not to the time of utterance of any sentence.

However, the question arises about the issue of phrase *takes place* — how it should be interpreted? For the states of affairs can also be qualified as past, present and future, even from our position in time. Therefore, it would be about the place in a certain time order, set by consecutive values of a time parameter. The world can be perceived as a history of consecutive states of affairs[47]. Given state of affairs is taking place if situated somewhere in the history of the world. The whole history of the world can be divided into certain sections, called temporal parts of the world, or the states of the world (in short S_i, where $i \in I$, or to the adequately adjusted to the cardinality of a set of such sections of a set of numbers). Of course the states of the world are also organised in time. The individual states of affairs can be considered as elements of given states of the world, that are certain total states of affairs in a given time.

In the light said above, the history of the world H can be described as a set of e.g. forms $\{S_0, S_1, \ldots, S_n\}$[48]. In turn, stating that *given state of affairs* — e.g. s_2 — *belongs to* H, may mean that there is such S_2 that $S_2 \in H$ and s_2 belong to S_2. Therefore the state of affairs s_2 in some sense is a part of the temporal parts of the world S_2.

The history of the world can be presented graphically by means of an axis, where the value of the time parameter increases from left to right, and individual states of affairs are located in the corresponding time values. Following this line of thought, I will reformulate (31) as follows:

Definition 2.3.1 *For any simple sentence* $z \in \mathbf{S}_{\mathrm{TD}}$*:* $f(z)$ *is true if and only if into the history of time falls the state of affairs indicated by sentence z.*

Obviously, by the history of the world, we do not mean only what was, but also what will be, what is etc. Therefore, we look at the history from somewhat timeless perspective. This approach is purely theoretical and does not yet settle the way

47 Developing natural intuitions related to the ordinal character of time in relation to events, present both in colloquial and philosophical thinking about time. For example, Broad wrote that *all events in the world history take their places in singular chain of moments* (Broad C. D. [39], p. 334).

48 This is only an example. I do not presume anything about setwise structure of time, nor about cardinality of a set of values for the time parameter, apart from its certain direction. It is possible to — and in fact I will — consider branching time (and its other possible variants); however, the individual states of affairs will be still situated in the parts of the history of the world, defined by the value of time, or its branches.

of judging such sentences, which state about future, although e.g. future does not exist or is not preordained. This problem will be tackled in due course. See the following figure presenting the history of some (lets take a chance that our) world, seen from atemporal perspective. Although the figure settles the existence of both a part of past and a part of future from present perspective. However, it should be considered also a visual aid only.

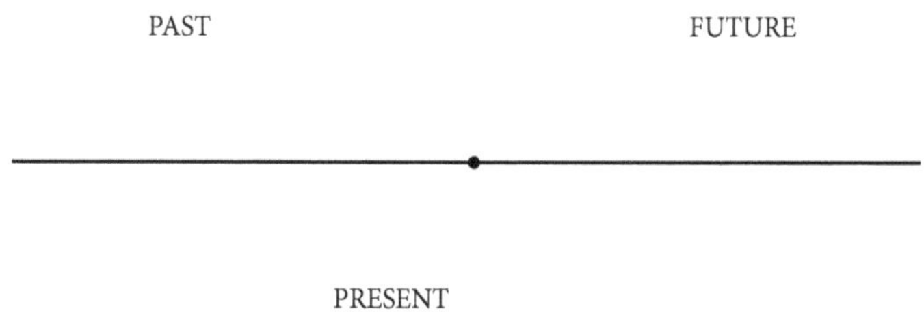

The above figure presents a part of history in the past, a fragment of present and a fragment of future. The concept of truth, discussed above, does not give precedence to any of these fragments. It does not state if the sentences about future carry own logical value in present or not. It only says that a sentence is true if and only if in time defined in the sentence described state of affairs takes place. Obviously, if there are no fragments of future (i.e. if the history of the world finishes in present, e.g. because it is not preordained how it will develop), then the sentences about future cannot be true.

According to definition 2.3.1, we can create a definition of false, by contradiction. Of course it requires the adoption of semantic law of excluded middle (or bivalence) which states that:

(32) *For any* $z \in \mathbf{S}_{\mathrm{TD}}$*:z it is true or z false.*

what through classical logic gives:

(33) *For any* $z \in \mathbf{S}_{\mathrm{TD}}$*:z is not true if and only if z is false.*

and through 2.3.1, the following fact:

Fact 2.3.2 *For any simple sentence* $z \in \mathbf{S}_{\mathrm{TD}}$*:* $f(z)$ *is false if and only if into the history of the world does not fall a state of affairs indicated by sentence z.*

Actually, this fact can be considered a definition of false. Fact 2.3.2 formed only by implementing (32). However, the use of semantic law of excluded middle encounters at least three objections. The first one is associated with determinist consequences, second, with existence of sentences that due to their objectlessness does not seem false nor true, and finally, third one resulting in general from questioning of this law, put forward on the basis of so-called semantic anti-realism.

The first objection shall be mentioned and elucidated now, while the other two will be the matter under discussion in further subchapters.

According to some philosophers, one of the sources of deterministic consequences **RDC** is the use of (32). On the one hand, it is said that we do not assume that sentences about future carry any specific logical value, but on the other hand, we claim that they are true or false[49] in any given time. But then any sentence $z \in \mathbf{S}_{\mathrm{TD}}$ must be true or false, and none carries any other intermediate value (e.g. the sentences about future could carry propounded by Jan Łukasiewicz value $\frac{1}{2}$ [168])[50].

Assume that we agree to the bivalence principle for sentences about future (at the moment, the case of past is not being considered)[51].

(34) *For any $z \in \mathbf{S}_{\mathrm{TD}}$, such that 'z' describes the present state of affairs: z is true or z is false.*

According to our previous considerations (31) a sentence stating something about the present state of affairs is false if in present the described state of affairs does not occur. The sentences stating about future time can certainly be characterised as a special case of sentences about present time, in a moment they state about.

However, the problem arises when we want to use the definition of false (fact 2.3.2.) for any earlier moments. What does it actually mean that *history of the world does not contain the state of affairs marked with z* (where z is such a sentence about future that says something about certain state of affairs s in moment t_2)? Does it mean that:

i) in moment t_2, s does not belong to the history of the world,

49 However, from the historical point of view, the use of this law to analyse **RDC** — shown in due course — seems desirable. Therefore I will try to champion it.

50 For as it has been accepted, logical values of true and false are usually represented respectively byl and 0.

51 Which constitutes a harmless assumption, although as I will explain in due course, not widely accepted — especially on the grounds of some so-called epistemic concepts of truth.

or does it mean that:

ii) in moment t_1 (such as t_1 is anterior to t_2), s does not belong to the history of the world?

The above distinction is crucial for propounding the third logical value. In fact, it comes down to understanding the meaning of *the history of the world* and its appropriate status, however, it affects the strategy of ascribing logical values to sentences. Let us try to face those questions sequentially, referring to both cases. Starting from the first one, as it is less general.

In moment t_2, the appropriate temporal part containing all state of affairs belongs to the history of the world. If state s is missing, the sentence z is false. The sentence is false also if *H it breaks* in one moment and there is no appropriate temporal part (marked with index 2)[52]. It seems quite clear. However, the *anterior falseness*, in moment t_1 proves problematic. Thereby we proceed to the second case.

Various strategies are available, and they can be focused around three main conceptual oppositions. First one: determinism and indeterminism. It can be acknowledged that the history of the world is complete in any moment of time, i.e. proper s-es belonging to proper S-es is preordained, therefore, the world is somehow *ready* both in the past and in future (e.g. the future states of affairs are unambiguously designated by present ones, just as present states by past ones). As a consequence, each sentence has a single constant logical value in each temporal part of the world. On the other hand, there is an opposite indeterministic standpoint. It would mean that not every sentence has a single constant logical value in each temporal part of the world.

The opposition to logical value issues is useful in describing the indeterministic standpoint. For logical value can be approached in classical way, with two logical values, or in non-classical way, allowing more than two of them. In the latter case, at least some questions about the future would adopt an

52 Depending on preferences, it can be said about lack of further temporal parts — it would mean absolute end of H and a flow of time; or it can be assumed that all later temporal parts are empty, meaning that time flows emptily. The temporal parts are though to correspond to the subsequent time points. For the semantic purposes, it seems that it would be better to adopt empty temporal parts. Then we know *where* the sentence is false. For in the absence of temporal part, the problem of *Plato's beard* occurs, as it is not known what the sentence relates to. In order to avoid this problem, we will offer a certain method of rephrasing sentences, so the condition of existence of an appropriate moment in time belongs to the content of the proposition.

intermediate logical value: would be neither true nor false. On the other hand, the two classic logical values could be used, while rejecting an assumption that there is only one history of the world. Allowing different alternative solutions, the sentence would carry a classic logical value, but depending on the scenario within which it would be considered.

The above approaches can be supplemented with another opposition. The opposition between the static and dynamic attitude towards logical values. Within static attitude towards a logical value, sentences carry constant logical value in every considered scenario. Particularly, when dealing with determinism, there is only one scenario of the future history of the world. The dynamics in valuating the sentences involves the ability of sentences to change logical values depending on perspectives in time. For example, sentences about certain future states of affairs are false until the state of affairs they describe occurs in the history of the world. However, if in the defined temporal part such a state of affairs does not exist, the sentence becomes false in a static sense. On the other hand, the sentences could be considered true, if there is every indication that what they state will happen, and their logical value becomes static only in the temporal part they state about (becoming e.g. forever false).

The dynamics of logical values could also be of the following nature. Sentence z is false in t_1 if in t_2 indicated a state of affairs does not occur, i.e. it does not exist in the history of the world, that is even earlier it had not occurred, sentence in t_1 therefore in a way *inherits* the falseness from the future and then it is always false[53].

In the case of indeterminism, regardless of the strategy, the history of the world would have been of dynamic nature. While the case of determinism is fully static in valuating. In fact, determinism deserves to be considered fully ontologically

---

53 Different interpretations e.g. **RDC** indicate such approach as correct in an attempt to recreate a deterministic livery of argument of Diodorus. In practice, while reconstructing formal aspect of the theory, one should adopt a formula stating that *if a sentence p is true now, it had been that p would be true*. This line of thought is also present in the deliberations far from the currently tackled issues. For example, Lemmon, while considering the problem of truthness of statements (in cited work he specified propositions, sentences and statements, appropriately conceptually differentiating them) he wrote that a statement expressed in a sentence *'Brutus killed Caesar.'* was true in any time, adding: *it was assumingly true even before 44 BC, when it could had been expressed by the sentence 'Brutus will kill Caesar.'*; in further part continuing: *we are coming close to the problem of sentences about future. All I can say is that 'Chè sera sera' is a logical truth, not a version of determinism (...)* (cf. [153], pp. 97–98).

deterministic — for it creates the strongest and indisputable foundations for unlimited bivalence principle.

The approach integrating the above opposition provides an opportunity to consider the competitive versions of the history, which emerge from the same past. Its advantage is that it also allows the application of the law of the excluded middle. When there is only one version of the history, it is determinism. The opposite situation represents indeterminism.

In the following chapter, we shall examine the possible ways of implementation of the conditions in question for the determinist case. Therefore, this chapter is devoted to the wide discussion on the question of determinism. The logical value of the sentences about the future requires though for the future to be in a way objectively established in present (cf. Łukasiewicz J. [167], p. 122). I may mean that such things as *the full causes* of all future events are present in already shaped history. Without them, there is no guarantee of the nature of future states of affairs, therethrough future history of the world may allow different alternative expansions. The line representing the future and present would then break in each *now*, further creating only numerous possibilities of which only certain ones will be updating along the passage of time.

Apparently, without additional assumptions, we are not able to give a full general concept of truthness or falseness of sentences. It does not mean that we will give up deliberations on the law of the excluded middle. Some arguments in favour of use of the law of the excluded middle in researches on **RDC** and against it shall be considered in the second part, in the subchapter on sentences about future, where we will present also the strategy of further works, accentuating the modelling of branches of history.

To sum up, the definition of false can be created in more natural way without adopting the principle of bivalence, simply adding it to our deliberations. However, if we allow the alternative histories, true or false of the sentence about future shall be viewed in relation to a given alternative future history of the world.

2.3.6 Presuppositions of sentences

The second objection regarding the bivalence principle stays close to the problems considered by Bertrand Russell within the theory of descriptions and arising from it an issue of presupposition [245, 246]. In his famous example, Russell examined the following sentence:

(35) *The present king of France is bald.*

This sentence, due to lack of such object as a present king of France, causes many problems when interpreted. One is that however valued true or false, it seems to state an existence of an object described in it by *present king of France.* Of course, Russell decided that sentence (35) could be rephrased as a conjunction of the following three sentences:

(36) *There is at least one such object as the present king of France.*
(37) *there is at most one such object,*
(38) *such object is bald.*

The above paraphrase created an opportunity to express sentence (35) in language **FOL** as follows:

(39) $\exists_x PKF(x) \wedge \forall_{x,y}(PKF(y) \wedge PKF(y) \rightarrow y = x) \wedge$
$\wedge \forall_z (PKF(z) \rightarrow B(z))$[54]

where — from the logical point of view — arises the formula:

(40) $\exists_x [PKF(x) \wedge \forall_y (PKF(y) \rightarrow y = x) \wedge B(x)].$

In sentence (40) (besides other conditions), certain presupposition of sentence (35) is explicated; it states that in *the universe* of reflections in general exists such object as *the present king of France.*

Similar problems, as in the sentence considered by Russell, may of course appear if sentences express propositions stating about other fragments of time, not only present. Some sentences may though not only assign certain characteristics to not existing objects, but even state that these objects possess certain characteristics in future (or even past!) time *t*, that will never exist in the history of the world. See the following example:

(41) *In 2045 Toruń will host Winter Olympics.*

It states that in year 2045 in the castle-town where Nicolaus Copernicus was born, sporting event called Winter Olympics will take place. In which case would we decide this sentence to be false? Is it only if in Toruń in year 2045 the Olympics will not take place? What if for any reason the idea of the Winter Olympics ceases to exist, or, worse still, the history of our world finishes before the year 2045? In such case, this sentence will prove objectless, and it would seem that also devoid of any logical value. For neither what it states happen nor its negation.

This problem may relate to sentences about different parts of the history of the world, in which the claimed states of affairs are supposed to happen. It needs

---

54 Where respectively $PKF(\ )$ — *to be present king of France* a $B(\ )$ — *to be bald.*

to be tackled in order to remain on the grounds of bivalent logic[55]. In order to eliminate this problem, we shall propose a certain general principle of understanding such statements, close to discussed considerations of Bertrand Russell.

Let us consider some examples of questions of obvious objectlessness. They relate consecutively to the past and the future:

(42) *Socrates in V BC watched television.*
(43) *Socrates will lead a discussion in 2050.*

Both sentences are determined in time — localise certain state of affairs in defined time frame. With the proper understanding of the individual phrases forming them, they also express certain propositions. In the standard meaning of appellation 'Socrates', it refers to a certain historical figure, important for the history of European philosophy and culture. The propositions expressed by (42), (43) state consecutively that Socrates watched television in *V* century before the Common Era, although the television is an invention of the end of previous millennium, and that Socrates, a historical figure, will lead a discussion in the middle of our century. Both propositions are therefore characterised by the states of affairs that had not or will not happen.

Thus it appears that these sentences cannot be neither true nor false, but rather deprived of logical value. Unless — following the Russell's example — we decide that the sentences carry proper presuppositions.

The accentuation of a preposition requires further insight into their logical structures. The following remarks of Frege from *Über Sinn und Bedeutung* shall prove helpful:

> Places, instants, stretches of time, are, logically considered, objects; hence the linguistic designation of a definite place, a definite instant, or a stretch of time is to be regarded as a proper name. Now adverbial clauses of place and time can be used for the construction of such a proper name in a manner similar to that which we have seen in the case of noun and adjective clauses. In the same way, expressions for concepts bringing in places, etc., can be constructed (Frege G. [76], p. 49).

In line with them, places, dates and time frames can be seen as objects, and their linguistic marks as appellations, e.g. individual or general names, and the latter as predicates. In relevant sentinental functions, they form true phrases

55 And not on the ground of multivalent logic, or logic with so-called gaps. The reason for constructing such logics is — among others — not only indicated problem of sentences about future, but also the lack of denotations for certain phrases used in sentences (cf. Priest G. [227], pp. 128–131).

for certain objects, and for others false (e.g. 'television — '(x)-*that is a television*' etc.). In order to make a point, the existence of presupposition in such sentences can be acknowledged even for individual appellations; that it is such element of domain which is identical to the denotation of an appellation (e.g. 'Socrates' — '$x = Socrates$')[56].

This approach enables translation of analysed sentences into language **FOL**. Then, the used examples of sentences shall adopt the following (and in this case similar) logical form.

(44) $\exists_{x,y,z}[x = s \wedge y = t \wedge P(z) \wedge R(x,z,y)]$[57].

If the sentences like (42), (43) will be understood in such way, then of course for any interpretation (where an interpretation means a model as in the model theory for **FOL**) they will carry logical values. They will be true, if: *i)* appropriate objects will belong to the domain, *ii)* those objects will be in indicated relations between them. However, if at least one of these conditions is not met, such sentences will be false

As a result, state of affairs should be considered under set theory. This issue shall not be elaborated. However, the evidence shows that when discussing states of affairs we mean objects in defined relations, in a certain time. The set theory

---

56 Admittedly, individual constant itself (e.g. s) should ensure belonging of given object under a domain in given interpretation. However, it may turn out that in time stated in the sentence, there is no domain that denotation s belongs under. (For we do not *select* a formula to represent given domain, but rather the domain in certain temporal part turns out to be as it is.) Then neither a sentence in a form '$A(s)$' (where A is a metavariable for any sentence where s occurs in), nor a sentence in a form '$\neg A(s)$' would be true. Another issue is of course defining constant s so each variable occurring in identity predicate with s denotes exactly one singularised object. The phrase '$x = s$' can be considered as definitionally equivalent to the phrase '$s(x) \wedge \forall_{y,z}(s(y) \wedge s(z) \rightarrow y = z)$', where the object denoted by variable x is said to have certain characteristic defining the object described in sentence s. This procedure guarantees singularity of possible object x, such that $x = s$. The issue of such object belonging to the domain becomes one of truth conditions imposed on the sentence. Of course a similar measure should be undertaken for all individual constants, leastways for those not determined as having denotations in a given domain. Intention to introduce an identity predicate shall become clear in the below examples.

57 Where individual extralogical symbols indicate respectively: t - individual constant denoting specific interval or point in time, s - person of Socrates, P - unary relation *being a television* or *being a discussion*, and R means ternary relation, depending on interpretation: *watching (by someone, something, sometime)* or *leading/doing (by someone, something, sometime)*.

approach to the empirical world is not an isolated one, for it is acknowledged in the literature[58]. The question of how far the ideas of the set theory can be applied to the empirical world that is variable and highly dynamic, we shall leave open[59]. We assume that it is a valid instrument leastways to demonstrate the existential presuppositions behind the analysed statements. The question of the general structure of reality the statements refer to is however another issue and strictly metaphysical problem. This issue shall not be discussed in this study[60].

Let us return to the question of presupposition. We have tried to indicate that presented sentences can be considered as assuming an existence (in fact *belonging* to described domain) of certain objects. However, not all the sentences in natural language can be transformed into so-called existential sentences. Many of them are of completely different structure. Let us see e.g. the following sentence:

58 For example, Andrzej Grzegorczyk writes: *Reality is: objects (things) of different qualities, connected by relations, forming various sets* ([92], p. 22).

59 There is also some criticism on this matter. For example, Adam Grobler expresses the following opinion, of which the second part relates directly to the discussed issue: *The very concept of 'correct type of a model' assumes that although the world's structure may not correspond to, commonly called, 'ontological commitments' of language, it does have a structure (accurately) describable in terms of individuals, functions and relations. Meanwhile, it cannot be excluded that the world, as it is, is a formless 'thing in itself', and if not formless, then made of something not resembling our individuals, functions and relations* ([88], p. 31, author's translation).

60 In the part on idea of *time* I will slightly focus on the issue of time ontology. It may shed some light on the issue of relations between states of affairs as static formations and time intervals in which they take place. It is worth to add that on the basis of specified answers to some outlined earlier uncertainties about the definition of *truth* and *false* and concept of *the history of the world*, in that sense of understanding of the sentences about future (and in general, analysed sentences) as presented in this chapter, interesting ontology based on the set theory can be developed. The temporal parts or states of the world can be seen as structures; states of affairs as their substructures. Then the history of the world would be a set of structures, and the time parameter would determine the structure in which given sentence would have been effective. Two basic problems to be solved are: in the first instance, the order within structures (depending on the adopted formal nature of time and in such general way not to settle the issue, and also the relation of time and states of affairs which shall be discussed in due course), and also an issue of sentences stating not about *punctual* states of affairs (discussed further), but about*ongoing* states of affairs and their variations. Such ontology shall not be presented herein. This idea, however it seems to be inspiring, is only indirectly related to the subject of this study.

(45) *One day, every Martian will be a chess master.*

In accordance with the previously discussed method, the logical form of (45) could be presented in the following manner:

(46) $\forall_x [M(x) \rightarrow \exists_y (t(y) \wedge MC(x,y))]$[61].

However, the evidence shows that the above sentence is true even if there are no Martians, as the above implication is then *emptily* satisfied. On the other hand, negation (45) is true when it is logically equivalent to the sentence:

(47) $\exists_x [M(x) \wedge \forall_y (\neg t(y) \vee \neg MS(x,y))]$.

It is equally disturbing consequence as the previous one. Sentence (47) is therefore true if there are any Martians in the domain, but there is not at all any moment in time, or in a given moment they are not chess masters. Nevertheless, we would like to determine that sentence (45) is also false if there are no Martians. There is a simple way to achieve it. It must be assumed that the general expressions also assume belongingness to the domain of the discussed objects. It should be determined that sentence (45) expresses information presented e.g. as follows:

(48) $\exists_x M(x) \wedge \forall_x [M(x) \rightarrow \exists_y (t(y) \wedge MS(x,y))]$.

In such understanding, (45) could not be true if Martians do not belong to the domain of discussion, and its falseness would be based i.e. on the very fact that they do not belong to the domain.

It is worth to mention that the definitions of *truthness* and *falseness* of the sentences in their presented interpretation based on a fragment of theory of descriptions and translation into formula **FOL**, do not pose a problem in relation to the so-called existential sentences containing a verb *to exist*[62]. Given that this verb should not be understood as *belonging under the domain*, but as *belonging under the domain of existing objects* (similarly, as we would understand *swimming* as *belonging under domain of objects that swim*). Therefore, it should

---

61 Where respectively $M(\,)$ - *be a Martian*, $MC(\,,\,)$ - *be a chess master in some time*, and $t(\,)$ - *be a time frame*.

62 For example the quoted definition of Aristotle from 1011*b* seems to apply only to existential sentences (cf. Woleński J. [294], p. 44).

not be considered as existential quantifier, at least not in relation to the synthetic sentences[63]. Let us consider e.g. the following sentence:

(49) *A philosopher exists.*

According to the proposed parametrising measure, the question should arise: when does a philosopher exist? For existence, just like every state of affairs, relates to a time localisation (and, as some may reasonably add, also a spatial one; we are, however, for obvious reasons interested essentially in the aspect of time. There were periods in history when philosophers did not exist. Let the sentence refer to 24th of July, 2003. Then statement (49) is equivalent, according to the expressed proposition, to the following sentence determined in time:

(50) *A philosopher exists on 24th of July 2003.*

can be rephrased in the above-proposed vein as follows:

(51) $\exists_{x,y}(P(x) \wedge y = t \wedge E(x,y))$[64].

Obviously, the negation of sentence (50) would adopt the following logical form:

(52) $\neg\exists_{x,y}(P(x) \wedge y = t \wedge E(x,y))$[65].

Let us sum up the above observations. In order to examine the sentences stating about states of affairs within bivalent logic, we need to assume at least that they presuppose certain issues of existential nature (in the meaning of belonging of the objects to given and considered domain)[66]. Often, these assumptions can be emphasised by analysing their structure from the viewpoint of **FOL**.

---

63 This function is sometimes fulfilled by a verb *to be*. It is though only one of its many functions (cf. Khan Ch. [132]). The relevance of this distinction can be observed in an imaginary example of entailment of sentence 'A man 100 meters tall exists.' from the additional premiss and one of the principles of **FOL**, in the form of '$A(a) \rightarrow \exists_x A(x)$' that is made in (Woleński, J. [295], p. 144).

64 Where respectively: $P(\,)$ is a predicate of *being a philosopher*, t is a constant denoting a date of 24th of July 2003, $E(\,,\,)$ is a predicate of *existing in a certain time*.

65 Not: $\exists_{x,y}(P(y) \wedge y = t \wedge \neg E(x,y))$, for the sentence would have been stating that in the domain, in an adequate time, there is an object of the quality of being a philosopher, but there is not. Personally I cannot see here any contradiction. As the philosopher could have been only an intended object, e.g. of purely literary fiction.

66 In any case it not being the only approach to the question of presupposition, although herein sufficient. The overview of modern approaches to the question of presupposition can be found in the work of Beaver D. [28].

On the other hand though, we do not claim that all the sentences of the type of our present interest can be well analysed in terms of their logical form within **FOL**. Moreover, we do not offer a formula for proper translation of any sentence. We postulate though that the assumption that expressions presuppose their own objective reference is a natural assumption and a forms proper, i.e. close to our linguistic practice, way to avoid expressions making statements about certain states of affairs in an objectless manner, and therefore surely not carrying either of the two logical values. I suggest lending the substance to the proximity of this approach to a regular linguistic practice with a means of an experiment involving reading following fragment and choosing one of the answers.

> Two neighbours are talking. One says: 'Mrs Kowalska, you should go to Szewska Street, because in the newly opened Social Insurance Institution they give extra money to people with the lowest pension.' Mrs Kowalska thanks and then goes to Szewska Street. To her surprise and disappointment, she learns that despite the enormous expansion and profligacy the institution raises its more and more modern buildings, there has not been one on Szewska Street built yet. Therefore, frustrated she heads back home. When she meets the neighbour, she says with a just reproach: (?)

Well, what does Mrs Kowalska say?

(*a*) *Your statement has not been neither true nor false.*
(*b*) *Your statement misled me. You have given me false information!*

I believe that Mrs Kowalska (if she is still on speaking terms with her neighbour) rather says something that expresses a meaning similar to (*b*). After all, she believes that the statement, that the said branch of the institution exists in reality, belongs to the content of information she received from her neighbour. For this is a regular presupposition in natural language, that while ascribing certain qualities to certain objects, we believe them to belong to certain domain. Stating in particular about objects from physical world, we assume their participation in the physical world, or existence.

In conclusion, we will add that the accepted here perspective of considering sentences is independent of **FOL**, what entails the following fact. In further analysis, we will be using propositional logic with suitable modal operators. These sentences shall be represented by means of sentential variables, without going into their internal structures. Therefore there is no practical requirement of translating any sentences into language **FOL**. However, as we can see, the question of interference between the objectless statements and the principle of bivalence can be solved by the requirement of adding to the statements the conditions of their objectiveness.

2.3.7 Sentences with modalities and other

Further comments will begin with an interesting fragment of the work of Marek Tokarz:

> (…) the most important laws of nature, always being idealisations, would become meaningless. As an example, let us consider one of Newton's laws, intentionally in quite archaic stylisation:
>
> *If a material object is not under the influence of nay force, the object moves with constant velocity in a straight line.*
>
> It has not been possible to find on the Earth nor in space such *material* object that is not under influence of *any* force. Nevertheless, Newton's law has a certain content and it is a weighty one. It does not state about our random part of the Space, but about all spaces possible to be thought about, even about those where certain objects are not under influence of any force ([282], p. 111, author's translation).

This passage expresses very important information about certain class of sentences, the so-called laws of science. These sentences do not state anything about any particular states of affairs nor about their occurrence in any time. They are invariant against time and space (cf. Such J. [265], p. 519). Among other things, it means that their logical value is *insensitive* to the time parameter, even though they undoubtedly belong to the set of constative sentences. Therefore, there are some reasons to, after the subtracting analytic sentences from set **Z**, exclude other types of sentences. These are such sentences which do not state any real states of affairs.

Tokarz indicated certain modal component of Newton's law. Obviously, there are plenty of statements that are *explicitly* modal. Let us discuss another four sentences:

(53) *It is allowed for a man to ride a horse.*
(54) *It is necessary, that Supremely Perfect Being exists.*
(55) *It is possible that hokey players took testosterone.*
(56) *Jan knows that Warsaw is the capital of Poland.*

These sentences from syntactic point of view should be considered as comprised of simple sentence and congruent sentence-forming functor of sentential argument. This functor is a certain modality (or a method: *modus*) operating on simple sentence. In the presented sentences, in their proper sense, the matter involves modalities, respectively: deontic, alethic, epistemic. There appears a problematic question if the sentences state about some states of affairs, stating their occurrence (or not) in our real world, or only state that e.g. certain states of affairs either occur in a possible world or occur in all possible worlds in respect to the initial world, as such modalities used to be interpreted within semantics of possible worlds.[67]

We are however interested in sentences of factual nature, i.e. stating something about occurrence of the states of affairs in a given time and in the considered world, not in various other possible worlds.

Some of the above sentences though do not need to have modal qualification characterised by the idea of *possible worlds*. Of course, it depends on proposition they express. For example, sentence (53) may simply state that *in a given time and country if someone is riding a horse does not break the law*. Similarly sentence (55) is allowed to be understood as carrying information that *certain hokey players in some time have testosterone* (hence they can use it, having it somehow *at hand*). The presented understanding of (53) and (55) is of course only one of many and claiming that each modal sentence can be *remodelled* into modeless ones without loss of sense, is undoubtedly false.

In further, formal reconstructions of **RDC**, the sentential letters will be ranging over primarily factual and simple sentences. The complex sentences will be constructed with the use of the classic logical functors (conjunctions $\wedge$, negations $\neg$ and others, defined by the former ones ($\vee$, $\rightarrow$, $\leftrightarrow$, etc.)). These functors are of course extensional (as opposed to modally understood functors from sentences (53)–(56)) which means that the logical value of the sentences featuring them is a function of logical value of simple sentences (while in modal sentences the matter involves intensional functors which use results in logical value of complex sentence not exclusively dependent on logical value of simple sentences. However, while analysing various sentences of the colloquial language, it has to be noted that it is easy to make a mistake in valuating a functor. See one of the most illustrative examples:

(57) *On 12th of September 1683 the king of Poland, Jan III Sobieski had arrived near Vienna and his army broke the siege.*

Sentence (57) is an example of sentences determined in time. Undoubtedly, from syntactic point of view, it consists of two simple phrases that with a certain (intended) meaning can be presented as follows:

(58) *On 12th of September 1683 the king of Poland, Jan III Sobieski had arrived near Vienna.*

---

67 For example, within the apparatus of modal logic, *it is allowed to interpret A* usually as *there is a world W′, available from W, that A in W′ is true*, but *knows that A* is interpreted as *for each W′ available from W, A is in W′ true.*

(59) *On 12th of September 1683 the army of Jan III Sobieski broke the siege.*

The main functor joining these sentences is 'and'. By *inattention*, we could therefore believe that (57) is a conjunction of two atomic sentences (58), (59) from the set $\mathbf{S}_{\mathrm{TD}}$, where the time parameter is adequately generalised into both sentences. It is true that it is a conjunction, but not classical, extensional. For in its intended meaning, this sentence is not *de facto* logically equivalent to sentence:

(60) *On 12th of September 1683 the army of the king of Poland, Jan III Sobieski had broken the siege and the king arrived near Vienna*

even though the regular conjunction is a symmetric operation:

(61) $A \wedge B \leftrightarrow B \wedge A.$

It is not equivalent because sentence (57) states that this state of affairs that the king *Jan III Sobieski had arrived near Vienna* has been in some sense (e.g. causal) the reason that *the army of the king Jan III Sobieski broke the siege.* Moreover, between both states of affairs, although it takes place on 12th of September 1683, there is a certain sequence of tenses. First *king Jan III Sobieski had arrived near Vienna*, and then *his army broke the siege.*

This example shows that sentence (57) within our theory cannot be considered as a complex sentence. For it does not include a classical functor in its structure. The modal component is included in its intentionality of meaning; as a result, this sentence does not simply indicate some state of affairs but also qualifies certain measure in which the states of affairs affect one another. Obviously, we could establish a form of the sentence that would be *less* modal e.g.

(62) *On 12th of September 1683, the fact that the king of Poland, Jan III Sobieski had arrived near Vienna, had been the reason (one of reasons) that his army broke the siege.*

But in fact not only states of affairs are being indicated but also their modal qualifications.

In the analysis of **RDC**, we will employ only sentences that — according to heuristically introduced sense of this conceptualisation — indicate or not the occurrence of some state of affairs. We assume that such sentences belong to a set of considered sentences **S**. For obvious reasons, we are not able to provide precise criterion of division of sentences of natural language into modal and non-modal (only modally qualifying state of affairs or indicating state of affairs), but only to signalise this problem[68].

In discussed further reconstructions, the complex sentences will be created with the use of extensional functors, but not only. For **RDC** in its essence is a *stricte* modal argument. For the thing is that it centres around certain temporal modalities. Such are i.a.: tensal functors (F — *It will be the case that...*, P — *It was the case that...*), or modalities *occurrence in given time* — $\mathcal{R}$, which we will use to build also complex sentences.

It needs to be stressed that the reasoning of Diodorus — discussed in due course — aims i.e. to define the modal functors *It is necessary to...* oraz *It is possible to...* in the very temporal terms, therefore characterises them in their temporal aspect. Therefore, these modalities, along with the above-mentioned ones, will be the only types of modalities we shall use. Other functors will only have extensional character. All of them will be thoroughly described in the second part of the book.

2.3.8 Truth, time vs. epistemic concepts

At the beginning of this section, discussing bivalence principle in relation to the statements about the future and past, we have highlighted objections from so-called epistemic concepts of truth. Now we will briefly discuss them, considering the trend associated with the so-called semantic anti-realism of Michael Dummett[69]. Within the semantic anti-realism, the unlimited use of the law of excluded middle (or bivalence principle) and the use of idea of truth with no regard to the cognitive access to the domain stated by qualified sentences as true (or false), are being questioned.

The epistemic concepts of truth emerge from the tradition of pragmatic and Kantian philosophies. Of course within the pragmatism, quite large theoretical differentiation dominated. In terms of the concept of truth, the creators of the pragmatism fluctuated between the idea of usefulness, as constitutive

---

68 For there is *multum* modalities. When asked: what can be modal, the author of book *Modality in Logic and Philosophy* answers, providing as many as eleven categories of different types objects, of which one, very extensive corresponds to sentences or propositions (Żegleń U. [306], p. 27). In turn, a comprehensive classification of modality concerning the sentential arguments and its explanation can be found in the work of Jerzy Perzanowski [214].

69 Omitting other epistemic concepts of truth e.g. concept of obviousness conceived by Descartes [57], and developed by Francis Brentano [38], or concept of cohorency (Neurath O. [201]).

for the truth, and the idea of consensus and approaching during the process of knowledge development, called *ideal truth* (James W. [117], Dewey J. [58], Peirce Ch. S. [212]).

The inspiration of different trends of pragmatism have led to arising criticism of non-epistemic concepts of truth, and also to formulating the epistemic concepts. This criticism can be found in various works referring to extremely different trends in philosophy: from postmodern philosophy to philosophy cultivated within analytic tradition[70].

Within the criticism of the non-epistemic concepts of truth, one trend can be distinguished, and it is called *anti-realism*. On its basis, some thinkers consider traditional dispute about cognition and object of cognition on linguistic, semantic grounds. Then the anti-realism takes form of semantic anti-realism. This attitude is understandable in the opposition to semantic realism, stating that:

> (...) the meanings of sentences (leastways constatives) become what makes those sentences true, or truth conditions. The above-mentioned conditions thus are perceived as occurring independently from our knowledge about them and ability to recognise them. As a consequence, the possibility of existence and phrasing sentences of which the truth conditions are beyond the reach of our cognition, should be allowed (Gutowski P., Szubka T. [93], p. 59, author's translation).

Thus the semantic realism stance accepts the thesis that many sentences of the language carry a certain logical value, despite the fact that nobody will ever be in such epistemic position to be able to establish this value. In the semantic realism, the key roles play two issues. Firstly, that the domain about which given sentence states, is (was or will be in given moment) independent of cognition. Secondly, that between a sentence (proposition) and what it refers to, there is a relation of such nature that the sentence is true (or false) regardless of whether any object performs an adequate act of cognition. These two characteristics cause the stance to be the realism, and to be semantic realism.

Questioning the two above postulates, anti-realism thus rests upon:

> (...) In essence this kind of anti-realism consists in the rejection of theories of linguistic meaning which identify semantic comprehension with understanding of the truth-conditions of statements where these conditions may transcend the recognitional capacities of speakers of the language. (...) there is general agreement among the authors mentioned that what is problematic about realist accounts of meaning is their common attachment to a notion of truth as objective conformity between, on the

---

70 In order to indicate the extent of the criticism, a list of the following works should suffice: Sellars W. [255], Putnam H. [232], Rorty R. [244].

> one hand, statements (and unuttered thoughts) and, on the other, objects and features in an independent, pre-structured world (Haldane J. [95], p. 17).

So anti-realists question the validity of considering sentenced as true or false without regard to the possibility of verifying them. In particular, this concerns the sentences stating about past states of affairs (not to mention the sentences about the future!) about which we can formulate plenty of statements, but very few can be analysed in any way in the context of their objective reference. One of the most famous anti-realists is Michael Dummett. He claims that, in the absence of appropriate evidence or traces from the past, we are not able to determine the logical value of many sentences within this class. Dummett suggests a criterion used in intuitionistic logic, where the recognition of a sentence entails indicating its effective proof. As long as the sentence is not proved it is not true nor false. Dummett's stance leads to the rejection of the principle of bivalence that states that each sentence carries one of these two values. This consequence, transposed into the domain of the synthetic, empiric sentences, results also in rejecting the unlimited use of the law of the excluded middle. The idea of truth is though replaced by Dummett by an idea of the effective decidability (cf. Dummett M. [66], McDowell J. [182]).

The previously discussed definition of truth has been however conceived as a non-epistemic definition. For we are interested in various sentences along with their logical values regardless of the cognitive access to any fragments of the history of the world. Such interpreted idea of truth is certainly not consistent with the anti-realistic approach. The following remarks are not so much to defend the outlined approach, and the more to account for a polemic with the anti-realism, but to explain some of its details. This summary provides a better opportunity for it.

The starting point for further discussions may be the observation that the proposed definition of truth does not constitute a criterion of truth. These questions are often distinguished in literature (Woleński J. [295], p. 175). This concept does not state the conditions required to be met for the sentence to be true, but what it means that given sentence is true. This point of view is practically consistent with one of the remarks of Alfred Tarski, expressing a cofeature of many non-epistemic concepts of truth:

> Anything can be achieved by means of substantially accurate for language learning definition of truth, one thing is unquestionable: this definition does not provide any criterion in each particular case useful for determining if in a given language a sentence is, or is not, true: moreover,

> this definition has not been intended to provide such criterion (Tarski A. [274], pp. 21–22, author's translation).

Since the concept of truth is not a criterion for truthness nor for falseness of sentences, in fact it can be discussed even if there were no true nor false sentences. What makes given sentences true or false is not only the presence or absence of the states of affairs indicated by these sentences. The fact that the sentence related to certain state of affairs is therefore related to the proposition it expresses, thus to the information transmitted. In order to emphasize this point, let us quote once more Wittgenstein.

> 4.024 To understand a sentence means to know what is the case if it is true (One can understand it, therefore, without knowing whether it is true.) (Wittgenstein L. [291]).

In this book, we are convinced that the sentences may be true or false regardless of the possibility to determine their logical values. The understanding of the proposition expressed by given sentence is, therefore, equivalent to the knowledge of the objective conditions under which it is true and — by negation — conditions under which it is false. Tadeusz Szubka calls such an approach *truth-conditional theory of meaning* and opposes it to the *verificational theory of meaning*, corresponding to the Dummett's ideas, where the understanding of the meaning of a sentence reduces to the knowledge of the conditions under which its adoption or rejection is justified ([267], p. 77 and others), not recognising its logical value.

According to the first approach, knowing a proposition expressed by following sentence:

(63) *On 20th of July 2003 it was snowing in Toruń.*

we do not know which circumstances should take place for the sentence to be true. However, because in Toruń, according to the calendar, on 20th of July 2003 it was not snowing, it is false[71].

But what if we consider a sentence from further past (e.g. from the time when there were no qualified dentists and part of their duties have been fulfilled by blacksmiths):

(64) *On 15th of January 1410 a tooth of Ziemowit from Pyrzyce is suffering from tooth decay.*

---

71 Obviously, the sentence could be understood in many ways, e.g. by checking in different calendar, according to which July is one of the winter months. In that situation, though, it would express different proposition than in the given meaning.

Even though from the anti-realistic point of view, this sentence is out of reach of the effective decidability and it cannot be assigned any of two logical values (not even decided to be true or false), within the classical approach[72] it meets the requirements of bivalence principle (even if it is seemingly objectless — considering previous propositions concerning presuppositions). Of course, its logical value is not known and it probably never will due to the absence of adequate cognitive access. Sentence (64) may however express a proposition very similar to the one expressed by the sentence stating about a suffering from tooth decay a tooth of a latter-day person, which we can verify/classify. The conditions imposed on its truthness in the past will in fact be not very different, except for the fact that we cannot refer to any direct medical opinion stating that Ziemowit's tooth is suffering from tooth decay — but it is not a condition for the proposition, therefore it does not result in sentence (64) not carrying logical value.

The problem whether the conditions imposed on the truthness of such sentences are (or will be) met, in fact more often than not surpass our cognitive possibilities. It would mean that there is simply a lot of *truths* that exceed our cognition[73], even if those truths are relativised to our method of verification of given statements and conditions currently imposed on them.

While applying such not limited by time idea of truth, problems related to reference of individual names such as 'Ziemowit from Pyrzyce' may arise. For in 1410 in Pyrzyce, there may have lived many Ziemowits. A proposition expressed in sentence (64) may state that Ziemowit in case is the one e.g. known from some given historical context. Then appellation 'Ziemowit from Pyrzyce' is an individual name. Sentence (64) may however express proposition stating about any person in the defined time bearing this name. Then the appellation 'Ziemowit from Pyrzyce' should be considered a general name of content *someone who bears a name Ziemowit from Pyrzyce* and believe that the sentence expresses existential proposition, that on 15th of January 1410 there is at least one such person called *Ziemowit from Pyrzyce whose tooth suffers from tooth decay.*

72 This approach refers not only to so-called the first Wittengstein but also to Frege, and nowadays developed by Davidson (Szubka T. [267], p. 137) and Polish school of logic and philosophy.

73 An interesting argument — known as *The Knowability Paradox* — in favour of the existence of such *truths* going beyond their cognitive possibility, developed by Fitch F. B. ([73], pp. 138–139). It is being perceived as an argument directed against the semantic anti-realism among others by Wiliamson T. ([290], s. 270–301). An outline of newer argument — according to its author based on Fitch's argument — is given by Künne W. ([143], pp. 167–169).

Similar measures can be undertaken in solving problems arising in the analysis of sentences about future states of affairs[74].

Obviously, there are many objects for which we cannot provide such individualising specification. We are also not able to provide it for many phenomena or states of affairs. For plenty of them irrecoverably escaped into the abyss of history, others will only make an appearance in future. However, it means only that from our position in time we are not able to express any propositions about them. But if possible in any circumstances, these propositions carry (at least in relation to past) one of two logical values.

This situation seems unsatisfactory. Travesting Wittgenstein one can say that the borders of our language are the borders of our description (true or false) of the world. It does not mean however, that from our position in time, future and past world can be fully described.

In the next chapter, we will focus on describing the objective conditions necessary to be met so even the propositions stating about future states of affairs could carry at least one of the two logical values.

74 The authors have adopted here various strategies in a situation called by Smart *naming future*, and in particular individuals, so as to be able to utter sentences about detailed states of affairs. One of them is presented by Smart, following Quine who proposes to dispose of individual names in favour of proper predicates and quantifiers assuring onlyness of such objects ([260]), pp. 93–94). This is an identical strategy as the one I have described in the subchapter on presuppositions.

3 Determinism

As the Biblical writer stated *There is a time for everything, and a season for every activity under the heavens* (Ecclesiastes 3, [277], p. 737). In the spur of the moment, one would agree that this statement seems to be suitable to act as a motto expressing the doctrine of determinism. A brief reflection brings, however, a thought that the term *determinism* is neither clear nor monosemous. For similarly as in the case of term *time*, there are many concepts of determinism (cf. Amsterdamski S. [10]).

In the previous chapter, while considering the problem of falseness of the sentences about the future, we have concluded that if sentences describing future would have to carry one determined logical value, then the states of affairs described in them would have to be objectively specified, determined. This means that the future part of the history of the world, just like the past one, would have to be in a way shaped. Let us use the following example:

(1) *On 30th of January 2020 the nuclear powers will forgo further production of nuclear warheads.*

(1) surely relates to future, as has been uttered on 18th of January 2012. If it is true today, than, according to the context of this statement and intentionality of meaning included in it, there will be such a temporal part, dated as 30th of January 2020 when the described state of affairs, i.e. resigning from further production of nuclear warheads by certain states, will take place. Regardless of the present and future states of affairs, truthness of (1) guarantees in 2020 the described by them course of events.

Such approach to the sentences about the future results in a stance that future history of the world has already been shaped. However, this question should be considered different from logical value of sentences about the future perspective. For the objective perspective determinates the logical values of sentences (apart from the above-mentioned semantic-linguistic issues). For it does not happen that e.g. the water freezes at the temperature of $0^o C$, because the sentence:

(2) *Water freezes at* $0^o C$.

is true. It is entirely the other way around. The sentence (2) is true because, in specified conditions, the water freezes in the temperature of $0^o C$.

Therefore, if the sentences about truth would have carried logical value, then objective conditions that will take place in future will have had to be already defined with some accuracy, preordained. Therefore, the future would have had to be in a way determined.

In further chapters, we will focus on discussion about such objective conditions that would guarantee logical value for any sentence, particularly for sentences about the future. First, we shall discuss in certain generality and different aspects the concept of *determinism*. Afterwards, we shall present its relation to the problematics and further conclusions and the strategy we will employ in the following chapters.

3.1 The extent of determinism

The sentences about the future (and sentences generally) can describe the events situated in different ranges of space, time and content. Therefore, they can characterise the world in various extents.

The sentences that can be assigned a logical value inform about certain states of affairs[75] situated in certain places and occurring in certain time e.g.

(3) *On 30th of September 1974 Jan Kowalski won the archery competition in Toruń.*

However, the sentence stating about Jan Kowalski's future sporting achievements could also be taken under consideration. Let it be the sentence:

(4) *Jan Kowalski on 30th of September 2020 will win sporting competition in Toruń*[76].

Sentence (4) is carrying a determined logical value, if it is preordained in any way that the person of Jan Kowalski shall or (shall not win the archery competition, which will have taken place on 30th of September 2010 in Toruń. This sentence thus carries the logical value at present, for the fact of the occurrence of the event has been already preordained. The situation is similar if the following sentence is considered:

---

75 In spite of the fact that within the definitions of truth and false I have been using the term *state of affairs*, here, and in further parts of the study, whilst discussing the objective conditions described by the sentences, I will be also using such words as *event*, *occurrence*. However, I am not giving them any specific meaning, but doing so only for stylistic reasons.

76 Of course while uttering the sentences about future events we will be using other grammatical tense than in relation to past ones. And yet, from the sentence determined in time interpretation perspective, important above all is the meaning of the word *win*, not the grammatical tense it has been used in. Further on, in the chapter on the pragmatics of the use of grammatical tenses, I will more widely discuss this matter.

(5) *On 30th of September 2020 Jan Kowalski will win the archery competition in Olsztyn.*

The above two examples indicate that for the sentences about the states of affairs that shall occur in the future, to carry logical value without imposing additional conditions on the places of the events these sentences state about, the question of occurrence or not occurrence of given events must be preordained for all the locations of these events. Naturally, it means that not only the question of currently existing possible locations of the events but also of those locations that over time occur in the world (e.g. cities that will have been built and in which the archery competitions with the participation of Jan Kowalski and other archers could have occurred) must be resolved.

A similar problem is concerned with the issue of the tense used in future sentences to describe future events. If, also in this matter, we do not superimpose additional restrictions, then the sentences describing future events, such as:

(6) *On 30th of September 3020 Jan Kowalski will win the archery competition in Olsztyn.*

will carry a logical value only if the events are preordained not only in any location in future but also in any time[77].

Another important issue concerning the sentences about the future is the question of their substantive extent. As we already know, sentences may describe virtually any spatial locations as well as any fragments of time. Nevertheless, the sentences can also, with different accuracy, describe the world — the content of the events itself. This of course relates also to the sentences about the future events. Let us examine another two following sentences:

(7) *On 30th of September, shooting his third arrow, Jan Kowalski will win the archery competition in Olsztyn.*
(8) *On 30th of September, shooting his arrow with a force of 200 N, Jan Kowalski will win the archery competition in Olsztyn.*

These sentences describe a certain state of affairs with different accuracy. If, similarly as in the case of spatial and time range, there are no additional conditions imposed on the sentences about the future, admitting that they carry determined logical value, the events they describe must be preordained at least in such extent as they can be described in. The question if the sentence describes

77 As well as if, in general, the fact of such time span is preordained.

adequately given state of affairs is of course above all else a linguistic problem, its understanding and verification conditions.

The above remarks can be summarised as follows: If the viewpoint that all logical sentences describing the future events carry logical value is to be accepted without any additional conditions from the objective perspective, then the future states of affairs must be preordained, determined in all the terms designated above.

This conclusion leads directly to the concept of the strictly determined future, relating to its all aspects.

3.2 Modal components of determinism

The fundamental issue occurring while defining the determinism is a differentiation between the determinism of theory and determinism of domain. Although these concepts differ (cf. Kałuszyńska E. [129], Wójcicki R. [297]), they can be mixed up[78]. For determinism in reference to the structure of reality is one thing, and the possibility to learn this structure, another, and — as writes Amsterdamski *recommendation on searching for the theories of certain types* ([10], p. 205, author's translation).

In the present chapter, we are however interested exclusively in the objective aspect of determinism. Consequently, we are focusing on the determinism in relation to the structure of the reality, or to the states of affairs.

Nowadays the objective determinism is often defined solely in relation to the physical world, excluding a wider spectrum. Though, the definition includes various modal terms. See the following two examples below:

1. John Earman gives the definition in Laplace's sense[79], in relation to the idea of physically possible worlds, i.e. worlds that are subjects of the same physical laws. Let W be a set of such worlds, then:

> A world $w' \in W$ is deterministic in Laplace's sense if and only if, for any $w'' \in W$ if w' and w'' are identical in a moment of time, then they are identical in any time (Earman J. [67], p. 13).

78 For example, statistical image of the world presented by modern science does not have to mean that the phenomena in the world are of statistical character. An example of this is the famous dispute on quantum mechanics between Copenhagen school and Einstein, de Broglie and Bohm on the nature of the microworld.

79 Of course, Laplace have moved the issue of determinism to the epistemological ground, as will be discussed in due course.

2. Determinism in a sense of David Lewis is also based on the idea of possible worlds. In addition, Lewis distinguishes so-called *diverging worlds*. These are the worlds of identical prior temporal parts, but they are not identical in further ones ([157], p. 359). On this ground, he defines the idea of *a set of deterministic laws of nature*:

   > A set of laws of nature can be called deterministic if and only if there are no two *diverging worlds*, subject to precisely one the same set of laws.
   >
   > On the basis of the above term, a conclusion of what it means that a world is deterministic can be drawn.
   >
   > The world can be called deterministic if and only if its set of laws of nature is deterministic (p. 360).

   This definition can be easily simplified by stating that a world is deterministic when there is no such a possible world which would be subject to the same laws of nature and which has identical prior temporal parts, but different further ones.

Both definitions implicate that the world is deterministic, because the existence of at least two possible scenarios of the future is excluded. In deterministic world all the future states of affairs are therefore determined in an unambiguous way excluding alternatives. Hence the definitions of determinism include the modal component also necessary in the definitions of indeterminism.

In due course, we shall more widely define the doctrine of determinism. Nevertheless, we shall not formulate any definition, but present different concepts of temporal parts of the world. Different, because generally — as Mario Bunge wrote — determinism is a certain ontological doctrine claiming that everything that happens, does it according to a certain preconceived plan, design, project (Bunge M. [40], p. 66, entry: *Determinism*). There are though at least two general conditions forming *differentia specifica* of the concept of ontological determinism, not only a condition of laws of nature.

3.3 Ontological determinism

The constatation that there is a certain order present in the world, called cosmos, can be considered as a historical base for the philosophical deliberations in general. In order to understand why the nature of the world is as it is, i.e. why certain types of phenomena are relatively frequent, while others — even though imaginable — do not occur in the world at all; or why certain types of phenomena precede and — as it seems — prequalify the ones of different types, but never the other way round etc., theoreticians began to search for laws that rule the propounded order.

The history of philosophy, and undoubtedly the history of sciences that actually largely constitute more mature continuation of many philosophical researches[80], can be perceived comprehensively, as the history of search for laws that rule various domains, thus as the history of attempts to identify rules creating the order of things in a given objective domain.

At the beginning of philosophical indagations, the theoreticians tried to grasp and describe accordingly the general, and in time more detailed, rules; it became a major objective for not only the philosophy, but most of all emerging exact sciences[81].

Assuming that all the phenomena in the world belong within a framework of order, i.e. that all are subject to sought-after regularities, and therefore preordained, determined and in a sense, necessary; we thereby express certain form of determinism. Depending on the strength of the order and its range, there are different versions of determinism, of which the strongest one states that no phenomenon — in the widest meaning of *phenomenon*[82] — in the world is not random, but determined by the order and its rules.

---

80 Well-known historian of science A. C. Crombie in his book *Medieval and Early Modern Science* ([51]) claims that natural sciences came from Greeks, writing that, among the first, they searched for a permanency that could be essentialised and generalised as a prototype of the scientific theory. Obviously, it is not a solitary belief. According to Werner J'ger, Homer himself — the ancestor of Greek philosophy — not only tells certain stories, but also seeks for reasons for the individual events ([116], note 34, p. 438). The dawn of philosophy used to be perceived as a certain genesis of subsequent science, and in the first ideas of Greek philosophers trying to explain and comprehend the world, not through the myths and religion but rather by means of theoretical reflection and observation of surrounding them environment. Both these cognitive methods, in different proportions, were also employed by various thinkers.

81 The twentieth-century science research, in spite of its advancement and theoretical and experimental sophistication, had not been much diverged from the presented purpose. Less than forty years ago, Rudolf Carnap described the laws of science as follows: *The observations we make in everyday life as well as the more systematic observations of science reveal certain repetitions or regularities in the world. Day always follows night; the seasons repeat themselves in the same order; fire always feel hot; objects fall when we drop them; and so on. The laws of science are nothing more than statements expressing these regularities as precisely as possible.* ([45], p. 11).

82 I.e. taking into account all properties of things, processes etc. that occur in the world of nature.

Among the historical theorists of widely understood deterministic orientation, from the ancient ones e.g. Leucippus should be mentioned (as one of the first he claimed that nothing happened at random, but everything from reason and by necessity (Diels H. [59], 67 B2), along with the continuators of his ideas: Democritus, Epicurus, Lucretius. Apart from the atomists, the world had been considered in deterministic categories also by stoics.

In modern philosophy[83], the continental philosophers of metaphysical uptake also stood for the determinism. For example, Baruch Spinoza in his *Etics* wrote: *nothing in the universe is contingent, but all things are conditioned to exist and operate in a particular manner by the necessity of the divine nature* ([262], prop. 29, p. 42), referring this view further, to the problem of contingency *...but a thing can in no respect be called contingent, save in relation to the imperfection of our knowledge* (note to prop. 33, p. 47).

The French materialists also expressed beliefs of deterministic nature. La Mettrie, the author of *Man a Machine*, inquired: *Why then would it be absurd to belive that there exist physical causes for which everything was made and to which the whole chain of this vast universe is so necessarily linked and subordinated that nothing that happens could not have happened...* ([144], p. 24) Holbach also have been rejecting the existence of a contingency. *there is neither chance, nor any thing fortuitous in this nature, where (...) all causes act necessarily according to fixed and certain laws* ([108], p. 39), and the very idea of chance: *no one effect in him, any more than in nature, produces itself by chance: chance (...) is a word void of sense* p. 103. Also David Hume, a philosopher absolutely far from materialism and advocating against the objective existence of cause-and-effect relationship as a necessary relation between a cause and an effect (cf. Krajewski W. [136], pp. 51–63), sympathised with materialistic ideas by writing: *there be no such thing as a chance in the world; our ignorance of the real cause of any event has the same influence on the understanding, and begets a like species of belief or opinion* ([111], p. 46). A criticism of the idea of the cause does not exclude that

83 Modern times provided many discoveries associated with the classical mechanics of Galileo and Newton that became the guiding light for other natural sciences. The laws of classical mechanics had been naturally presented in deterministic way (although some contemporary authors state that classical mechanics is not a deterministic theory, cf. Majewski Z. [172], Podgórski K. [223]). Therefore the fact that it helped to explain so many various and — it would appear — distant phenomena, like movement of celestial bodies and high tide, functioning of the circulatory or musculoskeletal systems in bodies of people and animals etc.; and predict many phenomena, later on verified by experiments, it allowed to think in a similar manner about the whole nature.

Hume spoke of *real cause. He just reluctantly regards the idea of necessary relation (…) [Apart from] all local distributional phenomena (…) in this regard it stays in full harmony with mechanistic beliefs of his century* (Krajewski W. [137], pp. 37–38) (author's translation).

The above-mentioned philosophers represent different periods of the history of philosophy, different schools and even different trends within these schools. Yet, they share describing the world in deterministic categories that indicates abilities to think in these categories in spite of differences in other matters.

These differences have their origin in another place. For giving a credit to deterministic means of arranging things allows asking more primary questions. The questions about the order in the world itself and its sources:

1. is the order in the world non-coincidental?
2. is the source of the order in the world located in the world or beyond it?
3. does the order of the world come under another order, under more general order?

The above questions may have different status (e.g. perceived as well/wrongly asked) and may be related to different answers. These answers depend mostly on the boundaries of the universe posited by researchers. And this can be at the very least dual.

3.3.1 Physical determinism

Some theoreticians accept a narrower universe of their considerations, restricting themselves to studying the order in the physical world and thus to cognition of the kingdom of nature. For these researchers believe that the most general questions are either void of sense[84], or there are no otherworldly factors causing order in the world which are cognisable — in actual fact by giving a negative answer to the majority of questions mentioned; if not considering them as *pseudoproblems*[85]. They believe that if yet proper questions can be asked about the cosmos and the order, then either it is the work of the blind chance, not showing any regularity and should not be a subject of science; or we do not hold cognitive access to the answers to the questions of such type. Therefore only the order in the world, the order that is not ontologically established in the cognitively

84 So-called metaphysical questions, condemned within different varieties of the positivistic philosophy of science.

85 Traditionally, the thinkers of both materialistic and scientific orientation of course do so.

available otherworldly order, remains the final subject of the research. As the world of nature is simply *physis*, such viewpoint can be called physicalism[86] in a broad meaning of the word.

The determinism in physical version states that *nihil sine causa*, so that nothing in the world happens without its cause in the world[87]. Even though the world itself and the rules that govern it could have been a work of chance, all that takes place in the world is not accidental, but determined by rules and by all that had taken place in the world beforehand.

The classic doctrine of determinism, known also as causal determinism, reads that no event is random, because its reasons affect it in a non-coincidental manner, i.e. in identical conditions they always cause the same effect. Ergo, every event is a physical consequence of certain conditions.

In modern times, the weight of the determinism have been transferred onto the laws of nature by postulating the nomological determinism that states that events are imperative because they are subject to strict laws of nature that are stable and unalterable (Krajewski W. [134], p. 86).

To sum up, the physical determinism is limited to the universe of physical matters, in its extreme version stating that, by the laws governing the whole world, every phenomenon is caused by different phenomena[88].

---

86 Applying the word *physics* in its quite primary and general meaning, as referring to the science of nature.

87 With different ways of understanding the term cause, depending on the favoured philosophy, which has its own historical exemplifications (cf. Krajewski W. [135]). Nowadays, there are various means of determining in a physical world described in science, not only causality *in the strict sense* but as to which description there is also no consent among philosophers of science. For example, in his work about causality, Mario Bunge lists at least six different ways of determining, different from strictly causal determination (Bunge M. [42], pp. 30–32), stating that: *to reduce determinism to causal determinism is to have either a poor opinion of the resources of nature and culture, or too high an opinion of philosophical theories* (p. 352). Moreover, Bunge defends this position consistently in his various other works (cf. e.g. Bunge M. [41], pp. 201–203).

88 Of course framing the issue of physical determinism through the laws of nature, one should take notice of the fact that modern scientific picture of the world of nature significantly differs from the presented approach. Up to the nineteenth century, the scientific discoveries had been providing the basics for the deterministic picture of the world, but further discoveries in various fields have changed this picture (e.g. findings in social sciences, biology and, most of all, physics: the kinetic molecular theory of gases of Maxwell J. C. [176] and Boltzmann L. [35], the statistical mechanics of Gibbs J. W. [82], and in twentieth century the discoveries within the physics of atom, initiated as far back as in nineteenth century by the discovery of

The doctrine of the physical determinism shall be presented by means of the following diagram. The symbols indicate as follows:

1. S_i — the total state of physical world of an index $i \in I$. The concept of total state of the world is similar to — although completely informally and only heuristically — to the term proposed by von Wright G.H. [300]. *The total state of the world* is situated in determined moment in time — it is also corresponding with the previously used term, temporal parts — and, as the name implies, is a state of all physical phenomena in this time. We do not preordain the number of the total states of the world, therefore let us assume only the linear order on I. The indices are assigned to individual total states of the world in such manner that if $i < j$, than S_i is earlier than S_j. Hence the sequence of the total states of the world will be called *the history of the world*, and its arbitrary, proper and complete subset (to be precise, a set X is complete if and only if: if $i < j < k \in I$ and S_i, $S_k \in X$, then $S_j \in X$) *a fragment of the history of the world.*
2. Index N means any but determined total state of the world called present[89].

---

the radioactive decay phenomenon and followed by emerging of the quantum mechanics with its main law — Schrödinger equation and Heisenberg principle, both, according to Copenhagen interpretation indicate to statisticality of laws of motion in the microworld) by accenting the statistical nature of many fundamental scientific laws, and consequently the statistical picture of the natural world. Noteworthy description of this methodological turning point can be found in the book of Władysław Krajewski *Konieczność przypadek, prawo statystyczne (Necessity, Chance, Statistical Law)* ([136], pp. 189–229). Essentially, it does not put paid to the possibility that the physical universe really is deterministic, and philosophical discussions on this subject — no doubt about it — are still ongoing. One of the interesting and most modern voices is the reasoning of Karl Popper in his work *The World of Propensities* and the previous one *The Open Universe. An Argument for Indeterminism* ([225]). In this work, Popper is straightforward about open-ended and indeterminated future ([224], pp. 28–33); it refers directly to the issues related to time.

89 It is clear to see that the order of the states of the world is not only the order of determination, but also of time sequence. For the total states of the world are determined in a given moment of time. However, I do not introduce time variable intentionally, as it would evoke additional problems (such as related to the question about the discreteness or density of time, and thereby of the set of the total states of the world). For the demonstrative reasons and for preserving the generality, I only assume that the total states of the world have different indices and that the order of the indices informs about the order of their succession and, at the same time, determination.

3. The arrows on the diagram indicate the direction of determination. They mean that under the laws of nature effective in the world and the prior total state of the world, the further state of the world takes place.

   Let us take into account such indices: j, k, N, l, m that:

$$\ldots < j < k < N < l < m < \ldots$$

According to the presented description, the following diagram corresponds to the states indexed in such way; the diagram can illustrate certain fragment of the history of the world if the subset of the indexes is complete.

$$\ldots \boxed{S_j} \longrightarrow \boxed{S_k} \longrightarrow \boxed{S_N} \longrightarrow \boxed{S_l} \longrightarrow \boxed{S_m} \ldots$$

State S_N, that is the present state, is directly determined by the previous state, i.e. S_k or such state S_i that $k > i >$ N, if state S_i exists. Since each state unambiguously designates the next one, then the correlation between the states is of functional nature. Composing correct functions allows to obtain unequivocal relationship between any prior state and the state S_N.

Such an approach to physical determinism proves that the present total state of the world in each prior state has been determined by the laws of nature. On the other hand, univocity also refers to the future states: therefore, they are as well determined considering the past states and the present one.

It is clear though that determinism in such total approach provides an objective opportunity for the sentences, describing events in any fragment of the history of the world, to carry logical value[90].

Although the presented above deliberations are only to be — as highlighted above — of heuristic value, they however cause multiple obvious doubts. One of them is the question of integrity of regularity and consistency of the laws of nature.

For, assuming that the world and the laws ruling it are the work of chance, there is no certainty that in one of the total states of the world they will not change

90 For there occurs a relationship, called in literature a nomological necessity, between the events. Of course it is reflected in the scientific explanation of the events, by referring to the laws of nature and other events. Interesting remarks on this subject and on differentiation within the discussion about the strict concept of scientific explanation (in contrast to the statistical explanation) makes Wesley C. Salmon in his work *Scientific Explanation and the Causal Structure of the World* ([251], pp. 15–21).

in effect of the external to the history of the world factors. On the other hand, there is also no guarantee that the next state of the world (in spite of the unequivocal determination) will have taken place in the history of the world at all. Therefore, in order to avoid similar problems, some of the theorists accepting the stance of physical determinism have moved the deterministic emphasis into a level other than physical, a metaphysical level[91].

3.3.2 Metaphysical determinism

The theoreticians accepting wider universe usually postulate that the kingdom of nature is only a particular fragment of the universe. The order revealing itself in the world is a manifestation of the greater plan and a number of reasons from outside the physical world[92].

Ergo, all the events in the world can occur according to the plan. In particular, this plan can derive from the Creator of the world, certain object distinguished from ontological universe, that created the world according to his design[93]. Therefore, there are some reasons standing behind the idea that the events in the world are of such and no other nature and that they are ruled by the determined laws of nature — these are the reasons of ontological, not physical nature. From the physical point of view, the world could have been different, but as it had come under broader, non-coincidental order, it cannot be different.

The above approach to the determinism would have also guaranteed a logical value for the sentences about the future. The recent drawing shall be of help illustrating it. For there would have been no changes in the world of nature.

Although the ontological determinism can be perceived differently. The fact that the nature of the events in the world is determined and that they are connected by some relations does not have to mean that in the world of nature

---

91 In search for the source and explanation for physical relations in ontological rationale (Krajewski W. [135], pp. 227–229).

92 A suggestion on distinction between the different physical causes and ontological reasons can be found in the work of Jerzy Perzanowski *Reasons and Causes* [216]. In this article, the author carries out a proof on the statement on the basis of certain assumptions on causality relation that states that the world consists of infinite number of facts and that this relation is not well founded. One of possible measures to avoid this statement is to refute the axiom stating that all physical facts are produced by other physical facts, and admitting that at least certain facts are caused by objects from outside the world.

93 For example, as the best of the possible worlds, just like in the theophilosophy of Leibniz.

there are any laws of nature at work[94]. For the sequence of events and their nature may have formed a part of a broader metaphysical plan.

From the cognitive point of view it would be allowed to describe certain regularities as laws operating in time, but from the ontological perspective there can only exist a substantive relation between individual total states of the world. However, the prior states cannot affect the posterior ones, as there is no physical relation between them[95].

Let us look at another picture illustrating this concept with symbols of the previous diagram:

$$\ldots \quad \boxed{S_j} \quad \boxed{S_k} \quad \boxed{S_N} \quad \boxed{S_l} \quad \boxed{S_m} \quad \ldots$$

The individual total states of the world are only related by time — or indeed by the temporal order, as they occur one after the other — and the substantive relation. Let us illustrate it with the following example: If in a certain total state of the world someone is an adult, then this person cannot appear as a child in one of the subsequent total states of the world[96]. And even though the individual

94 In the history of philosophy, a similar position has been presented by occasionalists (however, they considered mostly physical phenomena, not all of them). In their opinion the only real cause for events has been God, even though from the cognitive point of view it may seem that some things are causatively related to others. The fact that after the occurrence of an event of type *A* the event of type *B* always occurs, according to Nicolas Malebranche, has not meant that some *A* was a cause of *B*, but only that God makes them occur in this order [174]. Particularly inspiring has been the application of this doctrine to Cartesian mind/brain problem. The occasionalists, in an attempt to solve the problem of Cartesian dualism, indicated that wanting to move our bodies we are giving an opportunity for God, so he moves our bodies. By all means, an influence of mental aspects on the body has been omitted. And yet, from the empirical point of view, such a permanent correlation may seem like a regularity occurring according to some psychophysical law [173].

95 It is about the situation resembling a cine film. A cine film in time of projection provokes an impression of continuity of action. While watching a film one feels like the individual events were related and affecting each other, but in reality the scenes from individual film frame are only substantially related and there is no direct contact between them.

96 At least such a picture of the world, where things change naturally, not miraculously, draws up the rationalistic thought. However, if the world has been created by a malicious and omnipotent demon, an argument from Descartes? *First Meditation*

states do not affect one another in any physical way, the objective conditions for the logical values of sentences in any temporal parts are here guaranteed, as the events in the individual total states of the world have been determined and preordained in advance.

Both concepts of metaphysical determinism can be named predeterminism, as they emphasize the settlement of the history of the world and its complete scenario before it came into existence. From the divine perspective, the world happens within these concepts according to strictly defined ahead of time scenario.

It is worth to note that while drawing in this chapter a picture of metaphysical determinism as a result we have obtained determinism. However, the theory that the causes of the order in the world are located outside it does not preordain the determinism. For God would have created and defined the framework of the functioning of the world, while giving certain creatures a freedom to choose their ways of action, as in fact e.g. the Christians believe.

It is worth to note that while drawing, in this chapter, a picture of metaphysical determinism in this chapter, we obtained determinism as a result.

3.4 The consequences of determinism

All the above-presented concepts of determinism have been based on various factors designating the total states of the world. They can be considered as the concepts of global and strict determinism, i.e. which specifies all the states of affairs and their place in the history of the world. By all means, these concepts constitute nothing but some general frameworks, and the sole purpose of presenting them have been indicating such ontological visions of the world that guarantee a logical value for any sentences about the future. We shall not continue to differentiate them, and when considering the determinism, we will mean any concept of global and strict determinism.

The determinism leads to various interesting consequences. For the purpose of better discussion, some authors divide these consequences into various aspects of determinism, e.g. into ontological, metaphysical, logical and epistemic aspects (cf. Hankinson R. J. [97], pp. 518–520). For the time being, we shall hold on to a bit different classification suitable for further considerations.

([57], pp. 46–47), then individual total states of the world would have been surprisingly substantively inconsistent.

3.4.1 Logical determinism

The determinism surely results in each sentence from a set $\mathbf{S}_{\mathbf{TD}}$ having been given a logical value[97]. For if each state of affairs either belongs to or does not belong to a certain temporal part; therefore also, each proposition stating something about this temporal part (or about many temporal parts) is either true or false. Considering the whole set $\mathbf{S}_{\mathbf{TD}}$, there can be determined such its proper subset of true sentences $\mathbf{S}_{\mathbf{TD}}'$, that there is no true sentence that belongs to $\mathbf{S}_{\mathbf{TD}}$, and does not belong to $\mathbf{S}_{\mathbf{TD}}'$. Therefore we could have been able to point out a maximum set of true sentences in determined world, or, theoretically, its comprehensive description. It provides grounds to the use of principle of bivalence in the sentences about the future. Jan Łukasiewicz had introduced this principle in his famous reading, trying to demonstrate the idea of Aristotle argument from the considerations about the sea battle. However, Łukasiewicz noted that the argument of the principle of bivalence for clarity requires a support from another argument, of principle of causality (I shall not be introducing these arguments here, as they already form a part of classical and widely approved philosophical knowledge). This is because the sentences about the future require some objective foundation in the future, as the future events do not yet exist [167]. Tomasz Placek in his work expresses similar reservations to the logical value of the sentences about the future ([221], s. 23)[98]. In the above sections, we have not though been arguing in favour of determinism, but — as already indicated — we aimed to outline such objective conditions for all the temporal parts of the world to be determined as for every state of affairs that occurs in them. While in described conditions, sentences referred, in accordance to their values of time parameter, to corresponding temporal part, carry logical value. The problem of existence of the future states of affairs in the determined world is otherwise quite similar to the existence of the past states of affairs. The true sentences about the past also refer to non-existing states of affairs, e.g. the Battle of Grunwald, and yet their objective validity is not being questioned, as the Battle of Grunwald took place in corresponding temporal parts of a year 1410, continuing for a certain number of hours.

97 The term *logical determinism* has been introduced by M. Schlick [253].

98 On the other hand however, some authors advocate for the reality of the future; in their opinions, it results in the acceptance of the principle of bivalence for all the sentences, even though it does not necessarily lead to the determinism (cf. Faye J. [71], pp. 82–143). I shall present my viewpoint on this subject in further chapters.

3.4.2 Epistemological determinism

Obviously, the logical determinism must have an impact on cognition, because the result of the cognition is knowledge. Knowledge, on the other hand, expresses itself in organised sets of sentences.

The logical determinism specifies the extent of the possible knowledge about the world that can be expressed in the language. The knowledge in classical approach *(episteme)* differs from the ignorance and opinion *(doxa)*, by being undisputable[99]. The sentences that in the above terms belong to the scope of knowledge, must be true, or — in other words — the scope of knowledge cannot comprise the false sentences[100]. Notwithstanding whether knowledge would be seen as sentences about the individual facts or about relations between the entire classes of the facts[101], the logical determinism narrows down the set of the sentences that can belong to thus understood knowledge[102].

The above-mentioned situation is of course a certain idealisation. Nevertheless, it is reflected in the history of the philosophical thought. Pierre Simon Laplace had clearly illustrated it, introducing an omniscient demon hypothesis. Laplace, favouring the strict physical determinism, believed that a creature

99 In classical definition of knowledge, derived from Plato, there are three conditions expressed in the studies on the epistemic logics: *A person O knows that p if and only if i) believes that p, ii) , this belief is justified iii) a sentence p is true* (cf. Meyer J.-J. Ch. [189], p. 183).

100 However, this condition is characterised by a certain version of axiom T from the modal logic: $\mathsf{K}A \rightarrow A$ (where $\mathsf{K}A$ is read as *it is known that A*), which results in $\neg A \rightarrow \neg \mathsf{K}A$. $\neg A \rightarrow \neg \mathsf{K}A$

101 Of which proper part can be identified with the sentences expressing the empirical relationships, thus with the laws of nature.

102 It results in another interesting consequence. Some general sentences usually constitute generalisation of simple sentences, others are scientific laws, or sentences independent of time and space. Regardless of the source of determinism, either it would be the laws of nature or predeterminism, the general sentences, expressing some regularities independent of time-space location of the events, would have had a form of scientific laws, as if the universe would have been de facto a subject to certain laws. They would have stated that whenever the sentences of the type S_1, S_2,…,S_n, are true, then the sentence of a type S_{n+1} is also true (or true with a certain probability). Such general sentences correspond to the general sentences of the *explanans* within the nomological scheme of explanation (respectively: deductive and probabilistic) of Hempel-Oppenheim [103]. According to these laws, truthness of the sentences with *explanandum* could be explained in a common scientific way (cf. Bunge M. [43], pp. 3–133).

of absolute knowledge about physical universe in a given moment (about a location of the atoms and the forces acting on them) and about all the laws of nature (and even probably having an enormous calculative power!), could calculate physical state of the universe in any moment, both in the past and in the future, therefore *de facto* learn about the logical value of any sentence in the world (Laplace P. S. [145]).

Even if the world of nature has been determined, the only obviousness is that the position of Laplace's demon is not given to us[103]. By all means, it entails various problems concerning the cognitive abilities of the subject, its cognitive status and methodological restrictions etc. Although such position would never be available to a human, probably it is available to Laplace's demon or philosophers' God as one of his greatness.

3.4.3 Temporal determinism

Therefore, the world can be examined in a non-temporal manner, by considering the time as a set of indices, organising the successive events in a way we have proposed while introducing the term of ontological determinism. It is extra-temporal perspective, not distinguishing any index as corresponding to the current state[104]. In a derivative way, even the sentences can be considered within extra-temporal perspective, by referring to the described states according to their value of the time parameter.

In the empirical world, the temporal perspective is though constantly present. The sentence about the sea battle states about the future — from the perspective of Aristotle uttering it — events. The moment of the utterance of this sentence divides the history of the world into two parts: one, that has already happened, and another, that will happen[105].

Because we do not have an access to the omniscience, or extra-temporal perspective, from our point of view the majority of the sentences about the future seem not to carry any logical value, but at most some probability degree, and their value depends on further fate of the world. However, if the world had been of above-described deterministic structure, then regardless of our beliefs, the sentences, uttered in any moments of the history and expressing any events

---

103 And in the light of modern science, it cannot be given to us (cf. remarks of Elżbieta Kałuszyńska on this subject [127], p. 79 et al.).

104 It corresponds to *the arrangement of B* McTaggart (see next chapter).

105 The permanent change of perspective *past–present–future* is of course a part of general character of the time.

from any future states of the history of the world, would have carried logical value; it also applies to the sentence about the tomorrow's sea battle.

On the other hand, if any sentence at a given time turned out to be true, that would mean that in any preceding time it was true, and its negation was false, for its logical value has been already determined. Such opinion has been called determinism by Łukasiewicz ([167], p. 116).

Similar viewpoint, stating that any sentences carry logical value in each preceding moment of the history of the world, will call temporal determinism. Therefore, the temporal determinism in particular defines a viewpoint saying that the sentences stating about the future from the perspective of any, although determined, point in history, called the present, carry classical logical value.

While summarising this chapter, we shall outline the postulated relation between the temporal determinism and **RDC**.

3.4.4 Anthropological aspects of determinism

The reason why the determinism is such an important and permanent philosophical problem is probably the fact that, unlike many other philosophical issues, it has practical consequences in everyday life.

In the determined world, the human activity and creativity would have only been fictional. Playing poker or executing judicial punishment are examples of situations where the confidence in the determined world causes unpleasant cognitive dissonance. For the player, counts on the possibility to form the outcome of the game by means of his decisions and the prison system is based on the assumption that people act on their own will and are responsible for their actions. This is the very reason why the majority of philosophers argued against determinism.

One of the sources of considering the world in deterministic categories has been fatalism, a belief that certain events in the world are inevitable, and consequently, that human efforts directed to change course of events are fruitless. The fatalistic thinking motive often originated — although not only — from the religious attitudes, stemming from e.g. St. Augustine or protestant thinking (especially of calvinistic branch)[106]. For in these doctrines — as an essential element — there runs through a thread of predestination, accenting the belief that the Creator in advance divided people into damned and saved. Not

106 The fatalism usually denotes the determinism of mythological-religious provenance, where the forces from outside the world determine the human destiny. A good illustration of the mechanism of *fatum* can be found e.g. in the myth of Oedipus.

only the theological doctrines have to be the source of fatalism, but also the belief in a gift of precognition in some people, i.e. the ability to foresee the future.

The deterministic thinking about the world could have also originated from the philosophy of rationalistic and/or scientific orientation. Strictly following the rationalistic *credo*: *nihil sine causa* or *nihil sine ratione* leads respectively to the negation of the physical and ontological contingency.

The predetermination — unlike often fatalism — does not have to induce the indifference and activity suspension in the theory of human activity; quite the opposite — the ignorance of all the relationships between events, and as a consequence of logical values of majority of the sentences describing the future, present and past, allows subjects to act in such a way as if their activities were not restricted by any fixed structure of the world. The absence of complete knowledge about the world is a source of a sense of lack of a strict determination: *ignorantiae radix contingentiam est.*

This feeling creates a possibility to co-ordinate beliefs about freedom of action of an actor and determinism of the structure of the world. In the philosophy, a belief expressing this possibility used to be called compatibilism.

A representative of continental rationalistic thinking, Gottfried Wilhelm Leibniz's approach is an example of compatibilism. It is known that Leibniz's metaphysics has been based on the doctrine of the pre-established harmony, or the doctrine of predetermination. However, it had not precluded that the above-mentioned philosopher allowed freedom and responsibility of human activity in the compatibilitistic meaning. In some fragments, he wrote about it as follows:

> The entire debate concerning foreknowledge, fate, predestination, the end of life contributes nothing toward how we direct our lives ([152], p. 61).

> Hence, it is characteristic of one who loves God, i.e., the universal harmony, that he is content with events in the past. Because these things cannot now not have occurred it is certain that God willed them, and accordingly, they are best. But concerning future events, since no prior judgment is to any great degree evident to us, a place remains for the diligence of each with respect to deliberation and conscience ([152], p. 91).

> Thus since we know nothing of what is foreseen, we should do our part without pausing over the useless question as to whether success is foreseen or not, all the more since God is content with our good will when it is sincere and ardent ([151], p. 113).

The philosophers of compatibilitistic orientation represented various philosophical schools. Amongst the compatibilists, apart from Leibniz, there were Hobbes and Hume, and from philosophers less distant in time, e.g. G.E Moore. Even in the scientific philosophical studies, of e.g. R. Carnap, the compatibilism can be found. For in one of his books, he wrote that assuming the determinism, if someone decided to go to a musical concert, and we would have all the knowledge about the world, then *Was he compelled to go? No he went of his own free will. He is never freer, in fact, than when given a choice of this sort.* ([45], p. 219). For he acted of his own free will. Such an approach co-ordinates the determinism with the free will in a little different, more modern sense. The modern theorists interested in the issue of compatibilism believe that the main inconsistency between the compatibilism and an opposite position, i.e. incompatibilism, lies in the conditions imposed on the freedom of the subject and its agency (Honderich T. [109], pp. 111–123).

On the other hand, it is even being argued that the absence of any determinism does not allow to speak of the free will, for it is difficult to foresee events, and therefore make decisions. (Bunge M. [42], p. 133).

In the following chapters, where we shall present, among others, a historical material; we will mention that the Ancients perceived the consequences of **RDC** as aiming in a freedom of the human will[107].

3.5 Determinism *vs.* the Reasoning of Diodorus Cronus

In the previous chapters, we have described the general framework of the determinism that defines such objective conditions for any sentences to carry logical value. In consequence, the determinism is a sufficient condition for the belief we have named a temporal determinism. Therefore:

1. *If the world is determined, then it is also temporarily determined.*

The temporal determinism states that all the sentences about the future carry one of the two logical values. If the world is a subject to temporal determinism, then there exists only one correct description of the future. All that can be uttered about the future world can be expressed with the sentences from set **S**. As they carry a specified logical value, the set of all true sentences stating about the future constitutes a correct description of all the future states of the world that can be described from the present perspective. Therefore:

2. *If the world is temporarily determined, then from the language perspective only one correct description of the future exists.*

107 According to some researchers, it was the problem of the free will and various moral aspects related to it, that formed one of the most important reasons for the Ancient consideretions about the determinism (cf. Hankinson R. J. [97], pp. 513–514).

The above two theses result in the following theorem:

3. *If the world is determined, then from the language perspective only one correct description of the future exists.*

Therefrom, if the world is determined, then the various possible descriptions of the future are purely hypothetical, as only one of them describes the future in an adequate way. Thesis 3. is, however, logically equivalent to the following conclusion:

4. *If from the language perspective one correct description of the future does not exist, then the world is not determined.*

If therefore we are able to consider various alternative descriptions of the future and none of them is correct today, then the world is not determined. The future states of affairs are thus not unambiguously determined and only in the future it shall be decided which of them will have taken place.

When studying **RDC**, we shall take into account the theories allowing to formulate it. If given theory allows a possibility of alternative descriptions of the future, then according to thesis 4., it does not have deterministic consequences. However, if it does not allow alternative descriptions of the future, then along with one of the reasons of Diodorus, stating that the sentences about the past are necessary, it leads to a kind of *quasi*-determinism.

Quasi-determinism would have been stating that all the states of affairs that can be described by means of a set **S**, are determined in the history of the world. As for the other states of affairs (if there are such indescribable states of affairs at all) *quasi*-determinism has nothing to say.

The possible scenarios of the future will be further presented by means of branching structures of time, which allow different developments of the future, and thus, different possibilities. The necessary terms shall be introduced in the following chapters. Now we will focus on the third important issue, the time.

4 Time

Let us begin with one of the most famous quotes dedicated to the issue of time:

> What then is time? If no one ask me, I know; if I want to explain it to a questioner, I do not know. (St. Augustine [16], p. 242).

This fragment is actually not only devoted to time, but also to an attempt of its ideational conceptualisation. One could even risk a statement that in fact it is all about understanding the term *time* and an attempt to define it. According to some authors, the conceptualisation of time may be often confusing and could lead to many *nonsenses* (cf. Bunge M. [40], p. 297). What do we mean when saying e.g.:

(1) *When I was in plaster following an accident, the* ***time*** *flew so slow, as if it had taken all eternity; but as soon as the holidays started, I suddenly felt that half of it already passed.*
(2) *In culture x people perceived the* ***time*** *as a circle: it had neither beginning nor end, thus returning to a starting point.*
(3) ***Time*** *constitutes a form of intuition, meaning that, while perceiving phenomena, the subject imposes an order of consecution on them.*
(4) *In the Newtonian physics, two events are considered as concurrent if happen at the same* ***time****, but in the special theory of relativity, two events occur at the same* ***time*** *only considering a reference frame.*

In each of the above examples, in spite of the use of the same term *time*, it is about something different. Each use gives the term *time* a different meaning. Different, but showing common features. In the following subchapters, we shall discuss those different concepts of *time*. The presented differences shall cast some light on the answer St. Augustine was searching for. For the problem, he signalised arises — in my opinion — for two reasons.

The first one is associated with the very nature of the time, i.e. most of all with the fact that the time is not an object (in common, colloquial meaning of the term object) of a direct perception, such as trees, houses, street protests etc., but its nature is closer to the nature of abstract objects, e.g. mind, love or, at last, space.

The second reason is caused by the very chronological jumble of different meanings of the term *time*. Therefore, we could have asked St. Augustine — St. Augustine, do you know which type of time you are asking about? — suggesting that we can deal with the multiplicity of the types of time, corresponding to different meanings of this word.

In the further part of this study, we shall define the common ground, enabling prescinding from many inherent qualities of different concepts of time. Eventually, we will not be interested in the type of time we will be dealing with, but in its technical qualities.

In the summary of this chapter, we shall try to substantiate this approach and its application to the issues of the ancient reasoning.[108]

4.1 Cultural time

We live in a certain community. This community is structured in some way: has its own system of communication, created a number of institutions for solving problems, the global ones, as well as the ones concerning the whole community, or sectional, regarding individuals[109]. These institutions are of different nature. Some are public and distinct, such as army or school, others, though, are secret and, for unperceptive observers, absent while considered as commonly effective for every person regardless of their social affiliation, and therefore reportedly *natural* and almost virtually inbuilt in our biology, e.g. handshake, army bullying, clinking glasses while making a toast or euphoria when our national team wins the world cup.

Such mechanisms often do not serve any purpose, being relics of the past, a remembrance of rankling problems from long ago. However, sometimes their functions change — they relocate, in other times they fulfil many duties, of which some are open, and others, hidden. They may also fulfil — hiddenly or openly — the concernment of some groups of interest, which for this reason sustain them; or be of integrating function.

All these forms comprise the culture of the community, that is the manner in which it had gradually organised — consciously or not — its life in more or less hostile world of nature, probably governed by such purposes as survival and reproduction. This entire cultural baggage is concurrently carried forward. It is

108 Despite many differences appearing while considering different concepts of time, I am willing to consider the existence of one foundation for all of them — a feature which must be included in every concept of time. This opinion, I have described in an article *Różne oblicza czasu i ich wspó lny fundament*, [124] (*Different faces of time and its common foundation*).

109 Of course the word *institution* is used here in much broader sense than it functions in the colloquial vocabulary. It is about one of its technical meanings, present in social science, which allows to shed proper light on a number of phenomena remaining in the shadows when speaking about institutions in a regular meaning. An institution in the broader sense can be understood as *standardised and habitual line of conduct, constituting a foundation for social order in its fragment.*

not though completely stable and solid, but consists of dynamic components. The subsequent socialised individuals pass this baggage on to the next ones along with some modifications that occurred during their lifetime.

Therefore, it should not surprise that in such activities as an attempt to domesticate the nature for the community needs, there must have had occurred the need to some form of measuring the changes of defining distances between different practices. In other words, there existed internal need to organise activities within the community, based on the regularity and cycles present in nature. We should add: the regularities and cycles in nature crucial from the purposes of the whole community point of view. Such was the formation of a certain version of time useful in the organisation of one type of social life, but completely harmful in the case of another[110].

The issue of diversity of times within cultures seem, however, to be controversial among the cultural researchers and anthropologists. Some of them (Doob L. [62], Leach E. [148]) seem to suggest that in the analysis of given cultures, our Western measures should be employed to analyse time or to refer to it; others argue that the studies on the cultures different from ours (as exotic for us as e.g. Nuer people from North Africa, Tiv people from Nigeria or Hopi Tribe[111]) declare that feeling the changes in the outside world is related to how members of the culture experience the world in general (Evans-Pritchard E. E. [70], Bohannan P. [34]).

There is of course vast difference between the occurrence of changes in the physical world (or more generally speaking, in nature), regardless of the way it is perceived, isolated, classified, uttered and made use of by any human subjects, and the way these changes are perceived. Actually, we could place time somewhere within the first domain, the second identifying only with the perception of time. Then the problem of the cultural time would have been replaced with an issue of perceiving time in culture.

It would have been possible but for the fact that to the former we have no direct access. Only science provides the overall picture of the changes occurring in the nature. However, this picture is not really widespread. When asking a regular person about time and their own way of time organisation, extremely rare

---

110 According to some researchers: *every concept of time arises in the context of some human purpose and bears, inevitably and essentially the stamp of that human intent* (Lawrence N. [147], p. 24).

111 In whose languages *nota bene*, as Whorf's research were to prove, there are no grammatical tenses; therefore, *past*, *present* and *future* events cannot be described in such a simple way as in the languages, where the use of a proper tense is all that is needed (Whorf B. L., [289]).

in the first case, and never, in the second one; we would have been given an answer about the relationship between time and the reference frame, velocity of the observer, or gravitational mass. For in everyday life, we do not employ the concept of time present in e.g. physics (undoubtedly, such concepts also had not been in use by either St. Augustine, Diodorus Cronus, nor Aristotle — in spite of the fact they had been discussing time). If anything, it is closer to the Newtonian picture of time as flowing non-dependently regardless of the everyday events, but not necessarily at a constant velocity. Does this not mean that when we utter the word *time* on everyday basis, we think about something different than the time associated with the changes in nature? Is it not, being in fact an element of a common sense and colloquial language[112], largely a reflection of the culture we belong to? Something full of prejudices and settlements (substantially unscientific) about the nature of time, something aimed to effectively fulfil the tasks and requirements of the community we live in?

Perhaps, it would make sense to talk about the time in culture, not only about perceiving the time in culture, as, for the use of our lifetime activities, we have our own concept of time corresponding to our lifestyle, that, on top of that, may be antithetic to the scientific concept.

Some of the researchers, believing that it is worth doing, state that the cultural time is situated on a deep level of culture *consciously as well as unconsciously formulated, used, and patterned in different cultures* (Hall E. T. [96], p. 4). That is why in Hall's work — within the perspective giving the cultural time an autonomy — time is:

> (...) treated as a language, as a primary organizer for all activities, a synthesizer and integrator, a way of handling priorities and categorizing experience, a feedback mechanism for how things are going, a measuring rod against which competence, effort, and achievement are judged as well as a special message system revealing how people really feel about each other and whether or not they can get along. (p. 4).

As I have already mentioned, a hypothesis can be formulated that, not only due to the differences in the need of communities but also because of differences

---

112 Studies on common idea of the time are interesting and valuable, not only for broadly defined cultural sciences, but also from philosophical and logical points of view, and not only on the linguistic basis, leading to structuring the so-called tensile tense logics but also on the grounds of time ontology (cf. Snihur S. [261]). For it is an attempt to describe the time in which we live as a community. In further part of the study, I shall also indicate examples validating the thesis that one common concept of time, characteristic for all communities, does not exist.

in a natural environment development, in different cultures time can be of a completely different nature.

The concept of time ascribable to native population of Africa is more relatable to modern physical picture of time than the beliefs of the Europeans and North Americans. For many African tribes reduces time to events:

> In the primitive Africa, the events determine the time; they are not themselves entangled in its passage, but on the contrary — events tangle time in their own reality of the sequence of events; they not only organise time, but in a way also create it (Zajączkowski A. [301], p. 18).

A similar approach to the concept of time exists in the Ancient India:

> In the earliest experience, in Vedic India, time had been perceived as a proper existence of beings that we call sublunary. An empty time does not exist. There is only (chronological) spout of beings (...) (Panikkar R. [209], p. 112).

In the industrial countries, time has more instrumental character. Like everything in post-industrial world, time has its own price[113] and can be effectively processed:

> With AE peoples time is an empty container waiting to be filled; furthermore, the container moves along as if on a conveyor belt. If time is wasted the container on the belt slips by only partially filled and the fact that it is not full is noted. We are evaluated by how those containers look. If they are all full, that is a strong plus. If they are full with good deeds and creative productions, then we can feel we have lived a "full and productive life"! (Hall E.T. [96], pp. 78–79).

Doing nothing means wasting time. The model of time, associated with this vision, has a form similar to the picture of classical physics, that uses the concept of the absolute time that can flow *empty*, without events.

In some cultures, time had been perceived as a spiral or a circle, suggesting the return of certain parts of human history. The circularity of time has been assigned e.g. to the Hellenic World, contradistinguishing it to a linear understanding of time in the Christianity (where the beginning of the world: the act of creation, and the end of the world: the proposition day are strongly stressed) (Pàttaro G. [211], pp. 295–296). However, different authors claim that the concept of time in the Greek world is more complicated. Both periodic and linear concepts of time can be found there (Lloyd G. E. R. [160], p. 208), depending on what of the Greek culture tapestry is to be taken under consideration.

---

113 *Time is money* — we can hear this motto of Benjamin Franklin perpetually nowadays.

The differences in the conceptualisation of time in culture are probably associated with languages that given communities use. It is understood that in *the Classical Sanskrit the present tense dominates, and the future tense often adopts the meaning of the present tense* (Bäumer B. [27], p. 140). On the other hand, in the Semitic languages, the stronger emphasis is put on the action — if it is perfective or imperfective — than on the problem of the time this action takes place in (Gardet L. [79], pp. 154–155), which can cause such substantivisation of time, as in our culture. It is similar in Chinese language, where there is not so many complex tense relationships, as present in European languages, but in terms of tense it relies heavily on the context of the expression (Larre C. [146], pp. 83–84).

A similar and interesting problem poses the historical approach to time. Nevertheless, the visions of time in cultures — just like the cultures alone — are not fixed. This issue is correlated with the time measurement and means of its standardisation. Also the public demand have laid the ground of this issue[114].

The perception of time and its functioning in culture ultimately comes down to more basic facts:

> The experiencing of passage of time is a psychological fact, remaining in the complex, yet indisputable relations with the broader cultural context of which a person perceiving time is a participant (Zajączkowski A. [302], p. 6, author's translation).

It does not mean that the culture of community is reducible to the mental sphere of its individuals (the problem if it is so, or not, constitutes a certain discussed on the grounds of social sciences version of dispute, known from philosophy as the dispute between the realism and nominalism). It means that the cultural time is eventually experienced by some specific, conscious subjects.

Therefore, we shall briefly discuss the tense (not necessarily cultural) in reference to consciousness, not only to psychological consciousness, but also to pure consciousness, propounded by phenomenological philosophy.

---

114 In the medieval Europe, time had been indicated by means of bells, separately announcing the times for work, prayer, trade, or when for defence purposes, it was necessary to turn off the lights in town. Moreover, each town had own regulatory system for these matters, resulting in a certain degree of disorder. Only the introduction of new metrological measures around sixteenth century (the emergence of town clocks in the southwest Europe) constituted a source of universal order, while causing disturbance of *thinking and spiritual structures* of a medieval man. It constituted a transition into the modern approach to time in an universal manner. In 1370 in France, Charles V ordered all bells of Paris to adjust to the clock of the royal palace. This way, there had been created something we could call a national time of France (Le Goff J. [149], pp. 369–370).

4.2 Psychological and phenomenological time

During the trials and tribulations of everyday life, sometimes we happen to feel like the time was passing slower or faster than under normal circumstances.

Usually the phenomenon of *stretching* of time happens when we experience negative feelings or sensations, such as fear and pain. Time especially *drags* when we carry out an activity we find boring. The positive emotions, in turn, *compress* time. When carrying out nice or interesting activities, we can quickly notice that more *clock time* passed than it seems. Absorbing activities, fastening our attention on one, non-diversified activity, are of similar character and carry similar consequences. Performing these activities not only leads to *shortening* the time, but also to stopping it, *shrinking to nothingness.* This experience is present in e.g. surgical practice, when the doctors for many hours perform intensive operation and their attention is so preoccupied with their activity that they completely lose sense of passing of time.

All the above facts indicate that the very experiencing time and the way of experiencing it are different than the time associated with the processes out of consciousness of the experiencing subject. It does not mean of course that the experienced time is fully autonomic. To some extent, it can though take certain forms *sui generis,* corresponding only to given object and a nature of its experiences.

Following this path, time could be in general considered as a psychological category, related to the order given to individual experiences by the conscious subject. According to some authors, in a similar way James Clerke Maxwell was supposed to define time, stating that *time is, first of all, an idea — the idea that an ordered sequence can be recognised in our states of consciousness* (cf. Aveni A. [20], p. 5).

The emphasis on the role of consciousness in shaping the sense of time can be stressed in various ways. Sometimes, e.g. strong psychological clarification is attributed to St. Augustine's thought (Lloyd G. E. R. [160], p. 253). In Book *XI* of his *Confessions,* Augustine denies that time is related to natural phenomena. He states that even if the celestial bodies ceased to exist, time could be measured by means of man-made instruments (e.g. potter's wheel) ([16], pp. 273–276).

The Augustine's psychological approach to time[115] becomes most clear, when he writes that the present experience continues the history of the subject, i.e.

115 This approach is also interpreted as phenomenological (cf. Piłat R. [219], p. 75), i.e. related to the analysis of the subjective data of consciousness. Herein, I mean the same. When discussing the psychological time, I consider the perception of the passage of time that can be sometimes called psychological by the very reference to

the past exists now as a present idea or memory of the past event, and the future is present expectancy of the future events. Therefore, the present of the past, the present and the future things exists in mind. (p. 271–272). With Augustine, the matter involves characterising time, based on conscious thinking processes, remembering and psychological anticipation — expectations. Augustine successors, both in psychological and phenomenological grounds, stated — as shall be presented (*nomen omen*) in a minute — similar theses about the relationship between time and mind, or rather, the processes of consciousness.

Within modern psychology, the term *psychological time* is also used[116]. Within this conception, the issue of organising one's own experiences in the chronological order is only one of the components, however the reasonability to use the term *psychological time* requires such individual perspective. The perspectives,within which the time is not only one of methodological categories of human mind, which are *necessary for precise description of phenomena and their precise differentiation* (in an objective sense; cf. Pawełczyńska A. [210], p. 286), but is also understood as associated with the feeling of change, a sense of *becoming*, *passing*, and as well with a sense of dynamics of the statuses of the states of affairs from the present into the past and from the past into the present[117].

---

some *psyche* regardless of the changes in the outside world (such time, in particular implementation, can be different for different subjects). In this meaning, such perception can be and is studied by the psychology (more about it in due course). The issue of time on the *stricte* phenomenological grounds, therefore without reference to any specific psychological subject, shall be described in further parts of the study.

116 Although it consists of many aspects, such as among others:

1. the ability to chronologically localise oneself and the external events in time;
2. the ability of orientation in time without the use of instruments, such as clocks;
3. the ability to assess the passage of time, and also to determine the size of the time intervals;
4. the ability to localise oneself in the time perspective, localisation on the past-present-future line (Doliński D., Łukaszewski W. [61], p. 486).

117 This differentiation corresponds usually to McTaggartian approach to the two ways of recognition of time. One of them is to correspond to the categorisation of the states of affairs due to the relation of *being later*, while the second one is to be based on the change of categorisation of states of affairs from the present to the past, etc., what is to be, according to some authors, an important feature of perception of time in the natural cognitive systems (cf. Bielecki M.W. [32], p. 22). In the further parts of this book, I shall be employing the concept of time closer to the former one, although it does not preordain resignation from certain aspects of psychological time.

And for this, the mechanism of experiencing the present moment along with the experiences from the past and anticipating the future is needed (cf. Piłat R. [219], p.75).

The psychological time is therefore a string of experiences, a history of experiences (not a static one, but repeatedly anew qualified), it is organised and completed, and it has different aspects and various components. One of its aspects is the global dimension, e.g. associated with my memory of the events and its order since the day I am able to remember my first experience. Another aspect is of local nature: what events happened last year, in the last five days? The psychological time has also a metrological component: how long did my phone call last, has it taken longer than the morning walk with my dog? etc.[118]. Undoubtedly, such factors as: cultural baggage, external environment and the character of changes within the above mechanisms have a great impact on their operational mode. However, the psychological time is something different from the both combined. It is closely linked to the conscious mental life.

This fact has been probably best appreciated in the philosophical and psychological tradition which combined the concept of time with the concept of consciousness in the works of Franz Brentano [37], Henri Bergson [31], Wiliam James [118], and finally, on the grounds of the phenomenology in the works of Edmund Husserl [114].

These theoreticians believed that the answer to the question — why time flows? — requires the very studies on the essence of the consciousness. On the other hand however, also the consciousness cannot be understood as static, but rather through the reference to the metaphor of a stream or a river. Ultimately, both time and its passing, just like the consciousness, should be studied together. However, as the first three philosophers, whilst searching for the essence of consciousness, reached time; Edmund Husserl, whilst searching for the essence of time, reached the consciousness and its special structure as the necessary conditions of time (cf. Michalski K. [191], pp. 170–171).

According to Brentano, James and Bergson, the consciousness is understood as an actually indivisible whole. Similarly, its individual states also cannot

118 The majority of the mechanisms: locating oneself and the events in chronological perspective, qualifying events as present, past, future, the assessment of the intervals and orientation in time; is easily identified with the activities of certain parts of the brain, while examining people with dysfunctions of them (Damasio A. R. [55]). On the psychological level, however, these must be distinguished from the functioning of the biological clock, integrated with the body and regulating human activity in a way non-dependent from a realised time.

in fact be unambiguously isolated. For they refer to one another, without any clear borderline, creating a certain continuum. It is this construction of consciousness that allows the unity of the perceived objects (in different moments and various perspectives) and the unity of the consciousness perceiving them. Otherwise, there would have been a fragmentation, atomisation of experience that would have not been possible to put together without the continuity and the unity of consciousness.

According to James and Bergson, the unity of consciousness does not involve connecting different elements, e.g. present and prior moments. It is quite the opposite — in the present moment the past one is included, and also the future moment is emerging. This very division into points is derivative. The experience though is continuous and — as already mentioned — its substance is far from atomisation, therefore continuously experienced reality flows, forming a stream of time. Therefore temporality belongs to the essence of consciousness. It is not though the same temporality as in the objective world, where time is divided into units, and the states of the world are atomised. What is more, the possibility to capture the changes in the external world, e.g. the movement of the hands of a clock requires certain trace in memory, something that allows to determinate a change. The very proper concept of consciousness enables understanding the way the subject perceives this change. In the consciousness of the object, the past, the present and the future merge into a stream of experiences, into a river of internal time (Michalski K. [191], pp. 178–182).

Therefore, the reflections on consciousness lead to time. It is however the psychological time, not questioning the external time, even though emphasising its own independence. In the deliberations of Husserl, consciousness becomes a foundation for the existence of time in general.

Of course, Husserl's thought is not a part of psychological, but of phenomenological tradition. His research, Husserl conducted by means of the method of transcendental reduction and therefore in a mindset excluding all rationale and prejudging concerning the objective world, allowing only the sphere of phenomena. What is left, makes *the openly endless life of pure consciousness and, as its correlate on the noematic side, the meant world, purely as meant* (Husserl E. [113], p. 37). This method enables studying the way the phenomenon sphere manifests itself on the grounds of pure consciousness. Within the research on time, it means that it is not any objective time in question, but the time that manifests itself in the pure consciousness. On this ground, Husserl considers the problem of consciousness of the past, which he bases on the so-called retention (the consciousness of the future he calls correspondingly a protention). The consciousness primordially experiences some sensation (called a primal impression) which immediately becomes the retention and is maintained in consciousness, even though further appear, new primal impressions. Each impression occurs simultaneously along with the retentions of previous primal impressions. Hence, the stream of experiences superimposes, overbuilds of new impressions and retentions, creating the retentional consciousness. Of course, Husserl's experience of present moment, like in the psychology of stream, is not limited to an individual experience, but to the whole complex of primal impressions, retentions and protentions. The very retentional consciousness is a condition for consciousness of the sequence of time and a unity of objects in time (analogically to the psychology of stream), internal time, perception of external time, and also of itself by experiencing the retention. At the bottom of time and all these time objects, and their dynamics, there is consciousness (cf. Święcicka K. [269], pp. 69–73).

4.3 Physical time

The concept of psychological time used to differ from the concept of physical time. Within the consideraitons on physical time, there emerged three opinions: achronism, substantial concept of time (absolutism) and the attibutive concept of time (relationism) (Bunge M. [40], p. 297). As the first concept is a result of philosophical speculations, the two other constitute results of scientific reflection on the physical nature of the world. We shall now discuss some, crucial for further considerations, conclusions from these reflections, and the issue of achronism shall be addressed in the part about the philosophical issues of time.

4.3.1 From psychological time to physical time

St. Augustine, just after expressing his doubt about the knowledge about time, added:

> I dare affirm I know: that if nothing passed there would be no past time; if nothing were approaching, there would be no future time; if nothing were, there would be no present time ([16], p. 242).

Time is therefore to be linked in some way to a change of something (a quality in the world, a quality in the experiences of the subject). Sooner Aristotle noticed it in his *Physics*. However, accordingly to his doctrine, from within the framework of time, he isolated a material aspect related to motion, stating that the time was a *measure of the state of movement* ([12], IV, 221a, p. 110) and formal aspect related to a *soul counting movement*. According to Stagyrite, the formal aspect is as important for time as the one which embraces the movement in material world:

> Then, as to the former question, would there be any time if there were no living self? Without a being able to count, there would seem to be nothing countable, and therefore no number ([12], IV, 223a, p. 87).

The counting soul is very important for the existence of time, as without it, that is without a conscious subject, the approach to the matter of time only in a material aspect would, according to Aristotle, result in Ancient *aporia*:

> Next for discussion after the subjects mentioned is Time. The best plan will be to begin by working out the difficulties connected with it, making use of the current arguments. First, does it belong to the class of things that exist or to that of things that do not exist? Then secondly, what is its nature? To start, then: the following considerations would make one suspect that it either does not exist at all or barely, and in an obscure way. One part of it has been and is not, while the other is going to be and is not yet. Yet time — both infinite time and any time you like to take — is made up of these. One would naturally suppose that what is made up of things which do not exist could have no share in reality. Further, if a divisible thing is to exist, it is necessary that, when it exists, all or some of its parts must exist. But of time some parts have been, while others have to be, and no part of it is though it is divisible ([12], 217b –218a, p. 76).

The solution for this Aristotelian *aporia* is possible on the very basis of the concept of the active subject. It can result in extreme consequences, discussed in the quoted in earlier chapter strong interpretation of a stance of St. Augustine, which refers the future, the present and the past respectively to: memory, perception and anticipation, at the same time negating that the passage of time is related to the changes in nature.

The problem described above the two Ancient philosophers struggled with, and also indicated by Aristotle *aporia*, has been anticipated in the considerations of McTaggart on the issue of achronism, non-existence of time; where he introduces two series of the states of affairs. The arrangement of $A-series$ that divides the states of affairs into future, present and past ones, and arrangement of the $B-series$, which does not qualify the states of affairs in this manner, but only determines their order by relation of *being later than*. For the Ancients, time had been related mostly to the order of the first type. However, both series require changes, ergo a differentiation of the states of affairs. And although they introduce of course absolutely new perspective for describing the differentiation, as the McTaggart's division carries fundamental meaning for the time modelling in semantics of logic that we shall apply to the analysis of the reasoning; they essentially describe the same thing —change.

One of the types of all sorts of changes — if not the only one on its grounds — is movement. And the very movement have been the basic subject of modern physics and its further development, and the attempts to describe it resulted in the scientific concepts of space and time.

4.3.2 Time in scientific physics

We have written in *scientific physics* as for many centuries the philosophers had been developing a speculative physics. It is assumed that only Galileo Galilei on a broader scale introduced the experimental methods in physics and, of course, a mathematisation of its language, and quantitative approach, stating that the book of nature is written in the language of mathematics. However, it had not caused a shift from speculative methods in science. For it is safe to say that where there emerges a synthesis of knowledge, and a wide cognitive spectrum; the speculative element is always present in science.

4.3.3 Absolute time

We assume that the appearance of the work *Philosophie Naturalis Principia Mathematica* in 1687, written by Isaac Newton initiated the start of the era of empirical sciences. The method Newton employed was not only his own idea, but also a co-action of previous philosophers: Francis Bacon, Galileo Galilei and Descartes. As to the content of the works, there is even a dispute about the proportion of the contribution of Newton and Galileo Galilei in the discovery of the classical mechanics and mathematical-empirical method. Newton alone, differentiating philosophical elements from the empirical and mathematical ones, the former he did not value less, which may be suggested by the title of his work (cf. Heller M. [99], pp. 62–63, por. McMullin E. [183], p. 2).

In his work, Newton advocated for the existence of absolute motion in relation to the absolute space. However, he distinguished them from the relative motion and space, or motion and space in relation to human means of measurement and the accepted system of coordinates. The absolute motion of an object was taking place in an absolute space, measured by absolute time (thus similarly as in the case of previous elements, independent from any measure or measurement).

Time was absolute in the same sense as space, meaning that it was postulated as independent from the physical events. In theory, such time flew even in the world where there were no physical objects, and consequently, no physical change. Within the absolute time, the time interval has the same value (length)[119] regardless if at the beginning and at the end of it there are any events taking place. Therefore, two events are simultaneous, if the value of time interval between them equals zero. Thus, in the Newtonian physics, the idea of simultaneity has been as absolute as the concept of time.

In philosophy, the Newtonian concept of time is also called a substantial conception, given the postulated by Newton in *Principia* three components of the physical world: material substance, spatial substance and temporal substance[120], while none of these substances is reducible to the others, although, as can be seen, the time without events is possible, but the events are not possible without time and space. In relation to this conception, it used to be said that time and space form a certain container in which the laws of classic physics rule the movement of matter.

4.3.4 Relative time

The relation between the time and change is common and constitutive for almost all of the above-described non-physical concepts of time. The vision of time in modern physics confirms this characteristic.

The substantial concept of time and space accompanying the classical physics has been displaced along with the change of physical paradigm. It has been replaced by the attributive concept of time and space, related to the theory of relativity. Within this theory, time, space and their metric properties depend on the physical properties of the material world. The new concept of time is also a consequence of quantum mechanics, quantum theory of fields and cosmology, which provide essential information about time.

We shall however confine ourselves to the theory of relativity. From the historical point of view, the studies on the theory of relativity entail, among others, attempts to match the Newtonian theory to new findings. Establishing the velocity of light and discovery of electromagnetic waves have led to the belief that the interactions between the objects propagate with a finite velocity (not, as Galileo Galilei and Newton believed, with infinite velocity, that is *actio in distans*). The electromagnetic waves can propagate in a Newtonian empty space, as long as there is a medium to carry them. Such medium was called

---

119 I mean the difference between the end and the beginning of time interval in the metric approach to time. In classical physics, this value does not change regardless of the reference system.

120 The term *substance* is here understood in one of its traditional philosophical meanings, as a substrate, a building block.

the aether. The scientists have conducted a number of different experiments to demonstrate that for electromagnetic waves such medium exists. These experiments however led to mutually excluding conclusions, e.g. Fizeau experiment demonstrated that the aether is absolutely stationary, whilst Michelson-Morley experiment concluded that the aether was floating with objects (cf. Heller M., Lubański M., Ślaga Sz. W. [100], pp. 212–217, cf. Tempczyk M. [276], pp. 53–55).

The theory of relativity was a response to the problems. Within the theory, there are special (*STR*) and general (*GTR*) theories of relativity distinguished. *STR* is a particular case of *GTR*, describing the so-called inertial systems, that is systems moving relative to one another without acceleration.

Within *STR*, Einstein postulated two facts. Firstly, that the velocity of light is constant, and in consequence it is not possible to detect the influence of the aether on the propagating waves, and that in inertial systems all laws of physics are constant. The second fact — called the principle of relativity — constituted an extension of the analogical principle from the non-mechanical phenomena into all laws of physics. For the discovery of the electromagnetic waves suggested that on the basis of the laws of mechanics, by means of a measurement of the velocity of light in relation to some system of reference, it is possible to determine an absolute velocity of this system, which was in contradiction with various experiments.

An inertial system is usually understood as a system in a certain time and space that remains at rest or moves at a constant speed in a constant direction (Taylor E. F., Wheeler J. A. [275], p. 20). According to the principle of relativity, the laws of physics do not distinguish any such system and on their basis, within the system, we cannot indicate if it is in motion. The movement can be confirmed only in relation to another system. Therefore, the movement is a relative phenomenon and has no reference to any absolute system, such as stationary space. How have these studies affected the concept of time?

The physical research have given an answer to the problem of performing measurements of time and length within the same inertial system and determining the simultaneity of events[121], by measuring the distance between two observers. Events are simultaneous if they can be observed in the same moment from the midpoint of the distance between the observers. Time is then defined as the ensemble of the indications of similar clocks, at rest relatively to K, which register the same simultaneously (Einstein A. [68], p. 31).

---

121 In a space-time physics, the basic category, the reference of space-time parameters, is the very event (cf. Taylor E. F., Wheeler J. A. [275], p. 30).

However, the things get complicated if different reference systems are considered. It turns out, then, that the simultaneous events in one system can be not simultaneous in the other. What is more, the events preceding in one system may be observed in the reverse order in another reference system. This does not apply to the events that can be linked by a light signal. For it means that, in theory, these events could have been causatively related. Such events are observed in a fixed order in each system of reference (Katz R. [131], p. 35). We shall get back to it while discussing Minkowski cone.

The problem of simultaneity within different systems of reference results from co-called Lorentz transformation. These formulas describe the effect of the velocity of inertial systems on the basic physical parameters: length, mass and time. For time, Lorentz formula has the following form:

$$t_1 = t_0\sqrt{1-\frac{v^2}{c^2}}$$

where t_1 is a time variable in system U_1, which moves relative to system U_0, at rest, in which the time variable is of a value t_0, and v is a velocity of U_1, and c is a constant velocity of light. This formula defines the relation between the velocity and passage of time. The faster U_1 moves relative to U_0, the time in this system *flows slower*; this is called time dilation. The time dilation means that all the physical processes, including biological ones, occur slower in system U_0 than in U_1. Since the movement between the systems is relative, the Lorentz transformation describes the phenomenon of time deceleration in a symmetrical matter.

The time dilation phenomenon is usually illustrated as so-called the twin paradox and the father-son paradox. Let us discuss the latter one, as in fact they both lead to the same consequences, i.e. the rejection of a commonsense approach to the metrics of time.

Suppose that one man becomes a father at the age of 30 and at the same time, he also becomes a member of an international space expedition exploring the interplanetary space. At the moment of his departure, his son is 0 years old, as he has just been born. Let the spaceship, the father flies, be a system in motion relative to the Earth with such velocity that the physical processes occurring there happen ten times slower than on the Earth. One hundred years later, according to the time on the Earth, it turns out that the son is older than the father, as he is one hundred years old, while the father, who have been staying in a system where only ten years passed, is only 40.

The paradoxality of this example comes from the fact that it negates the equivalence of two commonsense definitions of *being older*. The first one states that among the living an older person is the one who has been born earlier.

While the second definition states that he is older who lived more life, that is, whose life has been temporally longer.

See the diagram below:

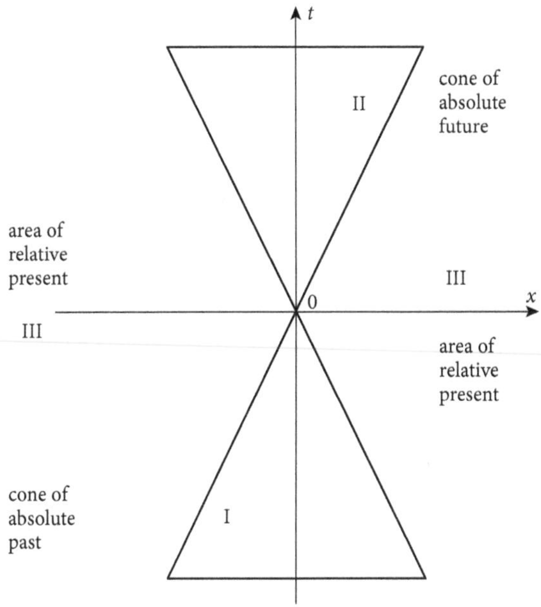

Another relevant difference in relation to the Newtonian physics is the fact that time and space within *STR* are not independent of each other, but form a wholeness called Minkowski spacetime.

The above drawing simplifies the situation as it presents the spacetime in two dimensions. Axis of abscissas (t) represents the passage of time, and axis of ordinates (x) represents the spatial distance. The connection point of the cones (on the drawing — triangles) is a zero point of some event that occurs in some present and in some time. The lines marking the upper and lower cone are so-called worldlines of the light. The worldlines mean a graph indicating a position of given object that is a function in time. The faster the object moves, the smaller the line inclination angle relative to the axis x. The optimum borderline is however set by the worldlines of the light that travel through spacetime with a constant velocity. Since all other objects move with a slower velocity than the light, the possible history of the world for point 0 lies between the fields designated by worldlines for the light.

The respective areas are of the following characteristics: the area of the absolute past (*I*) involves events earlier than the event zero, irrespectively of the reference system; the area of absolute future (*II*) involves events absolutely later than the event zero, irrespectively of the reference system, and the area of absolute present (*III*) involves such events that are simultaneous with the event zero in a certain reference system.

Given that the velocity of propagation of the physical interactions is finite and slower than the velocity of light, an event from point zero can be related by physical signal only to the events from areas *I* and *II* (the wholeness is called a causal area). It cannot though be related to events from area *III* because the difference in time between them is too small to enable physical interactions to cover the distance between these events. The possibility of physical interaction between two events thus indicates the locating of one of them in the area of the absolute past or future in relation to another.

The concept of space-time in *STR* does not consider the effects of the gravitational field. Only *GTR* takes account of the influence of the gravitational masses on the properties of time and space. Within this theory, it is postulated that the effects of gravity and accelerated motion are equivalent. The presence of gravitational field or non-inertiality of the system additionally slows down the passage of time, and thus all physical processes.

4.3.5 Properties of the physical time

The philosophers studying the physical time, facing certain problems involved in defining it, attempt to describe its features. Zdzisław Augustynek in his work *Własności czasu (The Properties of Time)* ([19]), lists the following features:[122]

1. one-dimensionality, which means in practice that each moment of time is assigned with one temporal coordinate;
2. continuity in a sense of real numbers; even though the continuity of time is questioned in the physics of the microworld, in philosophical literature it has been until recently believed that so far inefficiently (Such J. [264], p. 70);
3. non-density, that is, among others, open topology, corresponding to the model of a straight line (Augustynek Z. [19], p. 95);
4. unbranchedness — time does not split into branches in any moment;

122 These properties are also listed by other authors (cf. Mostiepanienko A. [197], pp. 41–45).

5. boundlessness; this property relates to the problem of boundlessness of the universe and depends on its different models, open topology suggests the boundless time; a straight line corresponds to it.

In addition to the above properties, Augustynek lists also so-called properties of symmetry, and amongst them: uniformity and isotropicity.

The time must be subject to the principle of uniformity stating that none of its fragments give precedence to any law of nature, i.e. that the laws of nature are invariant against time. In other words, but equivalently, there is no element in the structure of the physical time that would have been possible to distinguish from the others under the same laws of nature.

The issue of isotropicity, that is time reversibility, is however related to the question of the direction of its passage. Since this problem is of a great importance for the issues discussed within this book, we shall dedicate it a special part of the work.

An overview of different concepts of time, we have done, indicates their multirangeness, and often even exclusive relation. The time which we will model for the analysis of the reasoning, will not fulfil many of the above-mentioned conditions. For example, it will have an absolute metrics in each of its branches[123] (i.e. its metric shall not be in relation to the reference system, or simply the reference point will not be of interest to us); however, it would not necessarily have to be continuous or unbranched. The concept of time we shall be employing, will carry some constitutive characteristics and many variables that within the defined range will be possible to complement, obtaining a specific type of time.

Hereby, we shall discuss quite a problematic issue that associates the problem of time with semantics of the presented types of sentences, resulting in an issue of determining a logical value of the sentences in different time lengths. We shall return to this problem in a second part of the book, but in relation to definition of time provided there.

4.4 Time measurement, its accuracy and units

The connection of time and movement is present in every method of measuring time. As already mentioned, the first communities, while working out their own

123 Such terms as *branch* and *time* shall be precisely defined in the second part of the book, on the basis of previous findings of more philosophical nature. However, I would like to make a reservation that I will mean a certain special set ordered by specific relation symbolising the passage of time.

ways of measurements, relied on certain cycles and regular changes in nature, which affected their lifestyles. For many observable processes in nature are marked by regularity, both in macro sphere, that is cosmic, and in micro sphere, that is Earthly. These can be e.g. changes in positions of the celestial bodies, in particular the Sun and the Moon, ocean tides, seasonal animal migrations, or changes in the Earth insolation. Such phenomena, as long as based on an internal mechanism designating the passage of time and, in a certain degree, are regular, can be regarded as clocks. On the other hand, each attempt to pick up the rhythm of nature and describe it in the specified units, can be considered as creating a calendar, thus a way to measure time in a certain accuracy, including some extent of a margin of error and more or less accurate unit of measurement.

Initially, the calendars had taken partially mythic and partially literary form, while serving practical purposes, e.g. agricultural undertakings. A good example is a work *Works and Days* written by Hesiod that in a detailed manner links the astronomical phenomena with the ones occurring in the animate nature, indicating when certain rural works should be undertaken, such as planting crops, food production, animal breeding, etc. (cf. West M. L. [293], p. 381). However, in the course of history, the need for more accurate measurement increased, thus also for basing it on shorter regularities and more accurate units. This, in turn had a reciprocal effect on the other needs, inducing better organisation of time and improving measurement. The consequences were of course noticeable also at social and cultural levels[124].

Within many cultures and communities various clocks had been in use, and in consequence different calendars were created. These calendars, clocks and measurement units, due to margins of error, are co-definable also only within certain error tolerances. Obviously, the role of the basic time measuring method should be fulfilled by a system of the shortest units, allowing the smallest measuring errors. The most precise methods of measurement and scales are undoubtedly present in the scientific research.

The scientific research, depending on its subject, may however require the measurements of different time intervals. On the one hand though, science takes into account and measures large time intervals. These are used in research of e.g. modern geologists and astrophysics. The periods in question include the duration of such phenomena as *the evolution of the Solar System, of the Milky*

124 For example, one day the customers of courier posts began to demand their consignments to reach their destination within one day, regardless of the circumstances and distance, considering one day as quite a long period of time (cf. Gleick J. [83]).

Way galaxy, and even of the whole universe (Aveni A. [20], p. 113)[125]. On the other hand, within a microworld, there can be measured the time intervals of the length of an attosecond that is billion-billionths of a second. It is thus in theory the smallest unit of time in which the truthness of a sentence can be attempted to be verified. For a sentence stating about an event occurring in a shorter time interval is beyond the measuring verifiability.

The increase of the time measurement precision since seventh century (Chinese clocks) until now is from over one hundred seconds per day to 0.000 000 000 01 of a second (Aveni A. [20], p. 112). The history of time measurement is a history of inventing more precise clockworks and shorter cycles. It follows the path from the observable natural phenomena, through scaling them on sundials and water clocks, the invention and improvement of mechanical clocks, weight-driven and pendulum clocks, up to the clocks based on the cyclic phenomena of the microcosm. Nowadays the role of the most precise clocks is fulfilled by the atomic vibrations of caesium or other elements; on the basis of this phenomena, pattern of a second is being defined (cf. Dohrn-van Rossum [63], Mrugalski Z. [198], Zajdler L. [303]). At the same time, the replacement of measuring instruments induces a change and standardisation of units. Of course a large component of this change is conventional. We have divided a day into 24 hours, an hour into 60 minutes, a minute into 60 seconds, and the whole Earth globe into 24 time zones.

In reference to such precision, any longer time intervals can be defined, but it is not possible to measure them in more precise units without including a large margin of an error. The matter involves a certain conflict between clocks and scales. The conflict between clocks may however take various forms. It is already noticeable in the level of common comprehension of time in its medium scale. When changing the time zone, the travellers experience so-called *jet lag*, that is a phenomenon of discrepancies between own biological clock and given time zone local time.

While propounding a concept of time, we encounter a similar problem, though in a lesser extent. For the concept of time, we shall introduce, will uniformise the time and its unit, while the sentences determined in time have different units, which within given sentences can be harmonised, however often not without changing their sense. The concept of time we shall employ should allow defining the time units, referred to in these statements, by operating with a more basic unit.

---

125 Due to the radiometric methods, the astronomy and cosmology can reach into the very distant past, for there is where to reach out, as it is estimated that the universe we live in may be approximately a dozen or so billions of years.

Let us illustrate this problem, assuming that we are considering certain number of sentences stating about given year, e.g. 1995. That year many states of affairs occurred, among which there could have been also some that make one of the following sentences true (otherwise these sentences are false):

(5) *In July in Poland it rained.*
(6) *On 22nd of November in Spain a bottle of red wine was opened.*
(7) *On 5th of March at 22:03 in a fifth second in Nigeria a child was born.*

The model of time, where the above sentences can be assigned to given temporal parts, should operate at least as an exact unit as one second, considering bigger units as intervals composed of seconds.

But why our model should operate seconds, not e.g. nanoseconds (billionth parts of a second)? In theory, it is allowed to consider the logical value of a sentence:

(8) *In the 15th attosecond of July 1950 a caesium atom budged.*

The above doubts neccesitate the acceptance of such unit of time which shall allow arbitrarily small units in which there can occur states of affairs described in sentences from a set **S**. What units are these? — this, in the end, we do not know. For it depends on the nature of time and its construction. This issue shall be discussed in more detailed way in the second part of the work.

4.5 Philosophy of time and its problems

In addition to the issues concerning the time, isolated within scientific reflection, there is also a group of time concerning problems of *stricte* philosophical nature. In a large extent, the philosophy of time stems from a common understanding of time, providing justification to such beliefs as conviction about an unidirectionality of passage of time — its irreversibility, about a categorial difference between the past, the present and the future, and about a dynamics of time — about constant passage and becoming of reality. According to, involved in philosophy of process, David Ray Griffin, modern science, and, in particular, physics, does not provide any basis for these beliefs, most often considering time as one of the coordinates used for description of space-time location, where given event occurs. Naturally, it provides grounds for the philosophical considerations. However, the views of philosophers studying the time differ in terms of what attitude and autonomy can and should have the philosophical considerations and findings about the time in relation to physics (Griffin D. R. [87], p. 1–6).

The below presented philosophical problems about the time constitute only the tip of the tip of the iceberg of the philosophical considerations (cf. Le Poidevin R., MacBeath M. [155]). They have been selected intentionally. For the answer to the question constituting each of the problems has consequences for the time modeling, which shall be a subject of our further study. Therefore, these problems are not neutral for this issue. Presenting them one by one, we shall indicate which answer we are accepting and why. However, it will not be about an attempt to provide any solution, but about indicating certain inconvenience of the definition of time and embedded in it so-called silent assumptions — the assumptions deciding something about the nature of time.

4.5.1 Substantial time *vs.* attributive time

In one of the previous chapters, we indicated to the known scientific fact stating that the metric properties of time depend on different physical circumstances such as the reference systems, velocities or distribution of the gravitational forces. I added there that the considered concept of time shall not make the time metrics dependent from the reference system; thus being a non-relativistic concept. However, by postulating the absolute metric of time, does one at the same time postulate a non-dependence of time from the material world?

For within an attributive concept, the use of word *time* in its extreme form can be considered as brachylogy:

> Another example is a metaphor that time flows, which is a nonsense, as time is not an object. What 'flow' are real objects. Time is, lightly speaking, a velocity of real changes of the objects. (Thus time is not absolute, but relational.) However, the time is not a property of any distinguished object, just like space, time is common, that is shared by all objects. (More precisely, in relation to given reference system, time is shared by all objects which can be connected with this reference system by means of the electromagnetic signal.) (Bunge M. [40], p. 297).

The objectivity of time in given reference system does not mean its autonomy in this reference system; in ontological terms, it would be a primordiality of being. This is about something different:

> The physical time is usually regarded as objective, while the psychological time is, by definition, time (or rather continuance) perceived by a subject. The physical time is objective, but it does not exist by itself, regardless of everything else. The time is measured by observance of certain processes, e.g. swinging pendulum or radioactive decay. To be exact, time is not a subject to perception. We can only observe or experience certain processes. Therefore the person subjected to an experiment of sensory deprivation will soon have lost the perception of time (Bunge M. [40], p. 297).

The dispute on the ontologically objective nature of time is one of the most distinct philosophical threads in the philosophy of time. Newton's substantivalism towards the nature of time had been attacked by — each in his own way — Leibniz, Berkeley and finally Kant. Leibniz, while conducting the correspondental polemics with the Newtonist, Clarke, has been defending the relational concept of time and space, stating that:

> As for my own opinion, I have said more than once that I hold space to be something purely relative, as time is–that I hold it to be an order of coexistences, as time is an order of successions (Leibniz G. W. [152], p. 14).

The relationality of time in Leibniz concept did not however mean that the time was dependent on the reference system. Time is one and common for all events, but in contrast to the Newton's concept, without the consecution of things, the word *time* has no objective reference (cf. Heller M., Staruszkiewicz A. [102], Hartz G. A., Cover J. A. [53]). Therefore, Leibniz's metaphysical beliefs have not anticipated the relational concept of time in terms of the reference systems. What is also of importance, Leibniz dismissed the idea of time without things and events; meanwhile, in the general theory of relativity, the empty space-time is theoretically possible (cf. Heller M., Raine D. J. [101]).

Therefore, the dispute between Leibniz and Newton may be considered as concerning foremost the issue of ontological autonomy from physical world. Kant made an attempt to solve this dilemma in a creative way, by transferring the ontological problem, within his critique of reason, into cognitive ground. Kant claimed that time was one of the basic categories of cognition, allowing the external subject's experience to constitute itself. The idea of time is not being drawn from experience, but it is the very prerequisite of experiencing the world of phenomena in time. Kant justified his view by stating, inter alia, that the qualification of the experienced events as simultaneous or temporarily consecutive would not be possible if, on the ground of experience, there had been no idea of time. The philosopher from K'nigsberg concluded his comments as follows:

> (…) Time is a necessary presentation that underlie all intuitions. As regards appearances in general, we cannot annul time itself, though we can quite readily remove appearances from time. Hence time is given a priori. All actuality of appearances is possible only in time. Appearances, one and all, may go away; but time itself (as the universal condition of their possibility) cannot be annulled (Kant I. [126], A 31, p. 86).

Such an approach to the time and other cognitive categories allowed Kant's reply to Hume's doubts about the potential of mathematical natural sciences and the peremptoriness of his theses, despite the inductively limited database. These synthetic propositions *a priori* allow to formulate statements that are true regardless of the time and space reference. The dispute between Newton and Leibniz has found its solution in the first of antinomies of pure reason considered by Kant, which weighed two theses against one another: first one, that the world had a beginning in time, and was also limited in regard to space; and second thesis, that the world was non-finite in time and space (Kant I. [126], B 454, p. 458). Since time and space are only forms of sensible intuition, thus these antinomies are pointless ([126] A 525–528). For they assume the time independent from experience, like in the dispute between Leibniz and Newton.

The invention of non-Euclidean geometries had delivered some kind of a blow to Kant's considerations about pure forms of intuition — of time and space — named by him the transcendental aesthetics. The form of pure intuition has been excluding thinking of any other space than Euclidean space. However, the Euclidean geometry has been considered not only as a mathematical theory, but also as a physical theory describing the geometric characteristics of the universe we live in. It appeared however that there were other geometries possible, and yet they are, just like the geometry of Lobachevsky and Riemannian geometry, applicable to physical considerations, thus describing the physical world. These facts indicate that no external forms of cognition make the object think only in the Euclidean geometry categories.

Similar conclusions may be said about the time (cf. Szumilewicz I. [268], p. 11). The relativistic physics indicates that some properties of time depend on the properties of the physical world, which in turn do not depend on our perception. Metaphorically speaking, time does not flow everywhere in the same manner.

The controversy about the existence[126] of time, and rather spacetime, within modern physics, has its reference also in later philosophy, e.g. in the works of J. Earman, H. Field, where new arguments are put, and old ones are given a new look (cf. Gołosz J. [84], pp. 45–103).

Regarding an issue that is a subject of the dispute, in the considerations carried out herein, we will employ quite neutral definition of time. In certain alternatives, it shall however appear as the definition of time corresponding to the classical views of Newton. We will though not be interested if time is a certain abstract of changeable states of affairs, or if it has its own base of existence. The considered time may be obtained both by means of reduction to the state of affairs, or by considering it in an implicitly autonomic manner. The purpose of this study

126 I use term *existence* with a great reluctance because I prefer to believe in a traditional way that to exist means to be in time. However, such way of thinking about the time would lead to a paradox. It would be better then to say that time is. Although it sounds quite stiltedly.

is not the analysis of the physical time, but a very general and formal approach to time.

4.5.2 The direction of an arrow of time

The problem of the direction of passage of time, or in other words, meaningfulness of speaking about the direction of the passage of time, is one of the most interesting and the most frequently mentioned problem within the philosophy focused on the issue of time[127]. For example, above-cited Mario Bunge describes the problem of the arrow of time as follows:

> [The arrow of time is] an incorrect idea, that time flows from past to future. It is quite often claimed that the irreversible processes, such as the heat exchange, fluid mixing, ageing, universe expanding, reveal, and even define the arrow of time. This is an unfortunate metaphor, as the 'arrow', or directionality are proper for irreversible processes, but not for time (Bunge M. [40], p. 23).

Bunge's assessment results from the questioning of the absolute status of time. Even if guided by the concept of time reduced to the course of physical phenomena, we can still echo the Ancients that *tempus fugit.*

The daily experience, a ground for the life practice, indicates not only to the passage of time but also points to the fact that the passage of time is directed. The past events belong to the past, temporarily current to the present, and events that will occur, belong to the future, that is to something that — as noticed above *de facto* is not yet.

It is possible to say that the time flows from the past, through the present, to the future. It seems though like a very common-sense approach. For it is the time and its passage that differentiate the events into past, present and future ones. Therefore, we can write that in an everyday life, we experience a certain direction of time, its directional passage, technically called anisotropy.

The phenomenalistic approach to the time provides yet another important piece of information. The differentiation caused by a passage of time is asymmetric. The lack of symmetry between the past and the future is that we cannot change the past, but at the same time, we believe that we can affect the course and nature of the future events.

And yet we know that a common experience in terms of general nature of time is at variance with the scientific knowledge and thus in terms of physical time it is wrong. Is it similar with the issue of the time anisotropy? Whether in this subject

---

127 This section of the study is partly based on a previous article [123].

as well, the scientific knowledge undermines the common knowledge which we are willing to generalise to the time in general and within each scale?

Henryk Mehlberg in his work *Physical Laws and Time's Arrow* [185] wrote that even if the common expressions, such as: *here*, *there*, *up*, *down* specify physical space, we do not conclude, that the space is anisotropic. Similarly, the information implicit in natural language, referring to the time: *future*, *past*, *later*, *before* etc., should not constitute an argument for the passage of time having a distinguished direction. For the common knowledge on time, expressed in a natural language, cannot constitute a source of a justified belief about the directionality of the passage of time. Melhberg claimed that also the scientific time[128] did not have a direction. From the laws of science point of view, no direction of time passage is distinguished, thus there is no nomological criterion of the passage of time. The belief in the anisotropy of time would therefore be a common superstition, just like the belief about independence of time on the events. However, in science, the problem of anisotropy can be treated in two different ways, appropriately clarifying the very concept of anisotropy.

In the first place, there can be a question formulated if laws of science distinguish any direction of the passage of time, and, consequently, if there are any processes in nature that may occur only in one direction, or if all described phenomena in nature are symmetrical in time. This question is the very question about the nomological anisotropy. The negative answer to this question does not mean that, from the theoretical point of view, the physico-chemical processes — e.g. in a smoking cigar — would occur in reverse direction — and thus it would not be impossible that from a pile of ash, heat and products of decay after some time, an unburned cigar would come into existence. Many authors indicate that the phenomena, described by mechanical laws of motion, allow the passage of time to occur in both directions. For the laws of classical mechanics are symmetrical relative to time. However, there are phenomena related e.g. to the heat transfer that seem irreversible (Newton R. G. [202], p. 53).

If the conclusive answer to the problem of anisotropy cannot be found within the laws of nature, there is also a contingency of so-called conditional anisotropy, that is to base the arrow of time on the boundary conditions, the initial conditions of the physical processes. It means that, although the very laws do not prohibit going back in time, the conditions of their execution prevent such action or make it statistically improbable. Therefore, the conditions of the world we live

128 Scientific time is a time of which source of description is not a common experience, but natural sciences.

in would have been determining the direction of the passage of time. In other worlds of the same laws, but different initial conditions, time would not have had a direction. A statistical irreversibility of some physical and cosmic processes is known in science (e.g. entropy processes consisting in the energy potential equalisation and wave processes — based waves propagation from the source point).

However, the issue of anisotropy of time concerns all of the natural processes, rather than only their selected fragments; therefore, the processes or rules governing one direction of the passage of time in the whole universe should be indicated. Traditionally, in philosophy, this problem has been examined within three objective fields, respectively relating to causal, entropic and based on the cosmological models of the universe development theories of direction of time.

The concepts of causal direction of time, derived from, already classic, works of Weyl [288], Reichenbach [237], Carnap [48] and Mehlberg [186], (cf. Smart J. J. C [259]), made an attempt to link the passage of time with the causal interactions between events. The direction of the interactions is being identified with the direction of the passage of time according to the principle that the cause cannot be later than the effect. The causal structure of the world would have guaranteed the unidirectionality of the passage of time in the whole universe. However, the causal theories have been subjects to some strong criticism. For the very concept of causality is not clear and requires a good definition, wherein in its *definiens* there usually occurs a time sequence of the events, and this, in turn, is based on the concept of time with a direction. The analysis of the concept of causality is thus inevitable, as the causal relation cannot constitute a primary concept of an empirical theory. For it cannot be directly referred to the empiricism, e.g. the concept of mass or length (cf. Szumilewicz I. [268], pp. 47–48). Another important for these theories problem expresses a question whether actually the entire universe has a causal structure, which idea is hardly opposed in a modern picture of the microworld. The search for the criterion of the passage of time within the concepts seeking causes of the direction of time within the increase of entropy, may be considered as an attempt to avoid these problems.

The entropy theories of the direction of time refer to the second law of thermodynamics and the second law of thermodynamics discovered by S. Carnot and R. Clausius. The phenomena ruled by these laws tend to move from less probable states towards the more probable ones (Faye J., Scheffler U., Urchs M. [72], pp. 29–34.). Many researchers saw in them a possibility to indicate a nomological criterium of the anisotropy of time or at least a criterion for the passage of time in certain areas of the universe (cf. Schrödinger E. [256], Reichenbach H. [237], Grünbaum A. [89, 90, 91]). Without going into details, we shall only mention that the concepts of the direction of time based on the second law of thermodynamics, enable determining the direction of the passage of time as a statistically dominant direction of local changes of entropy. According to some philosophers, such solution is *satisfactory from the viewpoint of a materialistic concept*[129] *even though the criterion of the direction of time is not nomologically established and although we are not able to refer to any major universal laws, but only statistical distributions in the surrounding us parts of the universe* (Szumilewicz I. [268], p. 83, author's translation).

The problem of an empiric indicator of the direction of time have been attempted to be solved on the basis of the cosmological models of creation and universe expansion, attributing to the passage of time the direction resulting from the expansion of the universe (Szumilewicz I. [268])[130].

For the recent works on anisotropy of time present the problem from the perspective of the so-called block concept of time, with an attempt to suspend the phenomenological composite mentioned above (Price H. [226], pp. 3–21). This is connected to the modern attempts of providing arguments for the unreality of time which develop McTaggart's argumentation (cf. Mellor D. H. [188]).

Regardless of whether in the real world there is an anisotropy of time and its source lies within the laws of nature described by the laws of science, and therefore is of nomological character, or if its cause should be sought within the idiographic component, in initial and boundary conditions of studied phenomena; this problem is reflected in the modeling of time for the purposes of further analysis.

In our concept of time, we are going to exclude the possibility of the passage of time in an undefined direction. The time shall have a single determined direction. This decision can be justified as follows: In the **RDC**, as will be shown, the starting point is a differentiation of the future from the

---

129 That is a concept which considers time not as some non-dependent object, but as an attribute of changes on the grounds of matter.

130 A dispute about the arrow of time has not expired and is still present in philosophy of time. The direction of its passage is being consider also in relation to the radioactive decay phenomenon. Price H. [226]). Even though until the sixties, the time, in the nomological sense, has been considered isotropic (Such J. [264], p. 75), but the isotropicity of time has been compromised within the physics of, so-called, weak interactions. In 1964, American physicists, J. Cronin and W. Fitch, discovered a process of decay of neutral meson *K*, which depends on the direction of the passage of time. The processes of this type are not indicative of the nomological anisotropy of time on a global scale. For they are rare and occur in the microworld. Therefore, it is difficult to extrapolate them to the passage of time in the whole universe (cf. Nowikow I. [203], pp. 210–211).

present, corresponding to the Aristotelian doubt about the tomorrow's sea battle. This distinction indicates a clear lack of symmetry between the past and the future[131].

4.5.3 McTaggart's problematique

While discussing the concept of time, we have been, often quite imprecisely, employing such terms as: *time flows*, *past*, *present*, *earlier*, *later* etc. Certain allocation of these terms exist; it has been formed within the philosophy of time. This division is reflected not only in the concept of sentences, described above, but also within the concept of the states of affairs, as well as of the ways to represent time.

This differentiation has been described in McTaggart's reflections on the reality of time. In his already famous work, McTaggart tried to present an argument for the thesis that the time was real and substantial [184]. The outline of this argumentation can be presented as follows:

(McT.1) Reality of time requires the reality of change.
(McT.2) Reality of change requires a change in time qualification of the states of affairs (into past, present and future).
(McT.3) A change in time qualification of the states of affairs results in a contradiction (the same state of affairs has exclusive qualifications).
(McT.4) Therefore time is not real (cf. Lowe E.J. [162], pp. 62–63).

McTaggart's argument is not very precise. However, it does not affect the fact that there is an intense modern philosophical dispute about it taking place. Lowe E. J. [162], Le Poidevin R., Mellor D. H. [156], and Lowe E.J. [163]), in fact leads to a certain clarification of McTaggart's concept. We shall try to briefly explain his argumentation.

At the beginning of his reasoning, McTaggart introduced the following distinction: He stated that the description of the states of affairs, occurring in time,

131 Although, the logical value of the sentences has been considered as a criterion of the passage of time, i.e. if we believed that the time flew towards events not yet determined, which do not constitute truth-makers, then, at the end of the day, the **RDC** would have been breaking this asymmetry (Jarmużek T. [123]). In a different context of considerations, the same conclusions also reach other authors (Placek T. [220], Arsenijevic M. [15]), linking the issue of the passage of time to the problem of the determination of the universe. It is, in fact, a constant topic in the discussions between the theorists of the passage of time and the supporters of *the B-series*, that is McTaggart's approach (cf. Le Poidevin R., MacBeath M. [155]).

can be done in two different ways. The first one related to their permanent arrangement binary, transitive and asymmetric relation of *being earlier* (or *being later*). This arrangement is constant, for regardless of the time perspective, a position of the given state of affairs does not change in regard to the other ones. The second way of arrangement involves the qualification of the states of affairs as past, present and future, and depends, of course, on the time perspective, in which the arrangement is made. The above approaches to ordering the states of affairs in time are respectively called (*B-series*) and (*A-series*) (McTaggart [184], pp. 24–25).

McTaggart claimed that the reality of time entails the reality of change ($McT.1$), for change is a necessary condition of the reality of time. On the other hand, a change is linked to A-series, that is the qualification of the states of affairs as past, present and future ($McT.2$) (p. 26). The qualifications of the states of affairs as past, present and future are however in contradiction ($McT.3$) (p. 32), which leads to a conclusion (assuming of course the exclusion of such contrary qualifications) that there is no change and there is no such understood time ($McT.4$) (p. 34).

Not clear but interesting issue is an answer to the question on the correctness of McTaggart's reasoning, both the formal and material one. For on the one hand, this argumentation is criticised as faulty due to the misuse of causal phrases, used to describe the A-series (Lowe E. J [162], p. 65), and on the other hand, some of its rationales are rejected, thus supporting the thesis in a way (Mellor D. H. [188]), while other authors claim it to be completely correct (Horwich P. [110], p. 18)[132].

In accordance to the approaches to recognition of time highlighted by McTaggart, there are two ways of describing it and the sequence of the states of affairs. The first one, in more dynamic language, operating varying grammatical tenses, and the second one, more static, that uses predicates *later*, *earlier*, *simultaneously*[133]. These two approaches are even called two paradigms of the temporal logic, for they constitute an intuitive basis for two classic approaches to logic

132 The answer to this question is a separate issue, which I am going to discuss in another study. For in my opinion, McTaggart's argumentation alone requires more subtle and precise approach regarding such key terms as *change*, *time*, *past* etc. than are usually found in literature.

133 The distinction introduced by McTaggart is so significant for modern analytic philosophy of time that the debate over the reality of time and change, and the resultant deliberations about the past and the future have dominated its subject area (cf. Cockburn D. [49], pp. 3–7). It should be emphasised that McTaggart's stance in a significant manner fits into certain tradition initiated yet by Eleatic school

which takes account of the parameter of time. The supporters of these two paradigms are disputing over them, using many arguments for which one is substantially more accurate and more fertile (Øhstrøm P., Hasle P. F. V. [205], pp. 243–256)[134].

In our further reflections on the **RDC**, on linguistic and logic levels, we shall be employing some measures representing both a paradigm associated with the A-series as well as a paradigm associated with the B-series. In the second part of the study, we shall present the required for this purpose logic measures along with an operator R_i, which correspond to the B-series and logic with tense operators, developed within an approach corresponding to the A-series.

In fact, despite differences between both types of the temporal logics on the syntactic level, on the semantic level, where corresponding them point time structures are modelled, there exists — what is being noticed — an approach characteristic for the B-series (Hajnicz E. [94], p. 152). In such an approach, we shall prefer in, presented in due course, the concept of the structure of time.

4.6 Formal representation of time

We have looked at various uses of word time and different meanings associated with it. The purpose of this review was to justify the belief that there is no such thing as a single universally applicable concept of time, but there are many — more or less articulated — concepts, dependent on the reference point, perspective, range, culture and history. Of course, we are very far from claiming that the time is what everybody thinks it is. However, we maintain that there is no natural concept of time and therefore, while speaking of time, it is necessary to add in which perspective it is being described.

The above problem appears also during studying the reasoning of Diodorus Cronus and its important component, time component. For it seems that we need to decide which time Diodorus could have had in mind and which properties he could have assigned to it. We do not claim that Diodorus in his reasoning entangled the ideas of: time of experiencing subject, time designated by the calendar

---

of which views have been fiercely defended by Zeno, reducing to absurdity the existence of movement, and, therefore, of change and time related to the change.

134 In turn, within the deliberations on the metaphysics of time, the approach within A-series is perceived as more dynamic as well, while the approach within B-series is considered static and usually associated with the so-called approach of realism of the future within the philosophy of time. Newer arguments concerning these two opposing approaches within the metaphysics of time can be found in the collective work edited by Aleksandar Jokic and Quentin Smith [125].

of culture, time of biological clocks ticking in his body, time in its inertial reference system, or time in our part of the universe of quite random layout of gravitational masses. Presumably, however, the concept of time he did use could have been affected by some of the above-mentioned matters.

Unfortunately, there are not many testimonies on the views and philosophy of Diodorus left. In particular, we do not have precise knowledge about the concept of time he preferred[135]. That is why i.a. we are going to prescind from the material approach and address only its formal aspect. In defining the concept of time, we shall provide its general characteristics, leaving aside whether it is e.g. the physical time or the time of common thinking of the Greeks in question. The general character of this definition shall allow adding new conditions, which will result in obtaining a full formal spectrum of different variants of time, falling under a single general defined concept of time.

Apart from presenting various concepts of time, we have also indicated few problems belonging to the traditionally understood philosophy of time. Remarks we have made on this subject will enable defining the general concept of time in such a way that it shall correspond to some of our common ideas about the time. This is not the only advantage of the philosophical approach to the time. It will also allow for awareness that the concept we offer is not without philosophically engaged presuppositions[136].

In defining the concept of time for the purposes of further considerations, a compromise between generalisation of deliberations and its relation to intuitive and common understanding of the time is needed. Different alternatives of the time must allow a defined degree of generalisation, on the one side, and on the other, they should satisfy some commonly uncontroversial assumptions

---

135 However, there are certain evidences that he had been a supporter of the concept of discrete time. Some authors make of this fact to develop an interpretation and reconstruction of **RDC**. It will be mentioned in this book in the part about reconstruction.

136 Anyway, for some authors working within phenomenological tradition, the definition of the structure of time I shall provide will not be in any relevance to the description of individual experience of the time. It shall rather express other type of experience, constituting an abstraction of many individual experiences, which loses many characteristics of the former ones (cf. Ogrodnik B. [207], p. 19). On the other hand however, the mathematisation of time which will be close to us, according to many historians, through the ideas of Kant, has become an inspiration for different mathematical considerations, included e.g. in works of Hamilton (cf. Duda R. [65]). Thus there is a certain connection between the mathematical time and the experimental time.

about the time, e.g. that time has its direction. These assumptions shall also be partly confirmed in the historical material we will discuss.

The general concept of time will therefore express a formal-ontological approach to the time. This approach is called a formal since, as mentioned above, it does not prejudge much about the material aspect of time. It is also called ontological as it allows different possible alternatives. This approach is not about answering questions about what is time and what it is like, but what types of time are theoretically possible.

4.6.1 Attempts to define moments of time: moments as points *vs.* moments without points

Whilst defining the concept of time necessary for further considerations, we shall employ some tools from the set theory. Thus we will view the time as a structure, i.e. a set of specified objects, which constitutes the scope for certain order relation. Such concept of the time has of course some advantages and disadvantages. Amid weaknesses there is surely the fact that the concept of the structure of time we shall thereby obtain has in fact a lot in common with the time, e.g. the physical time as well as with any other objects of which structure is isomorphic to the defined concept of time. Such understood time is thus of purely formal and extensional nature.

Notwithstanding, this approach has also some benefits. These include: the highlighted above general character of studies and the availability of modelling various formal alternatives of time and undoubtedly the precision that many other philosophical considerations lack. The approach to the time through the idea of structure finally allows to engage to the issue of **RDC** appropriate logical tools, corresponding to the very different variants of the structures of time.

While defining the structure of time[137] in the above manner, we need to specify the domain of the relation. The elements of the domain, according to the established tradition, are usually called moments. From a purely formal point of view, we can consider however different candidatures for moments. This problem essentially stems from the attempts to define the time through abstraction, as characteristics of simultaneous events (cf. Ajdukiewicz [1], pp. 384–387). Regardless of the issue of simultaneity, that is whether it is considered in an absolute way or within a reference system, the question about the nature of the moments remains open. For if the

137 Or the time, as both these terms shall be used interchangeably from now on, meaning only formal aspect of defined object.

moment is considered as an abstraction class of simultaneous events, it inherits a certain feature from the events. It is possible, e.g., to operate the concept of events understood as non-dimensional points, then of course also moments should be seen as points. Some examples of such an approach are works of Zdzisław Augustynek ([17, 18]). On the other hand, there were concepts of moments without points, based on sets of simultaneous events, but events considered as extended processes (cf. e.g. [85]). The above concepts are associated with the mereological approach to the ontology[138].

As already signalised, we shall prescind from the material nature of time, excepting the problem of what are the moments and what is their relation to the world of the states of affairs. A reflection on the **RDC** shall be conducted within a pointwise concept of time. Such an approach enables the use of relational semantics, associated with the modal logics which constitute natural means to express and reconstruct the reasoning of Diodorus Cronus.

The possible alternatives of time shall be called logical structures of time, as they are linked to the specified logical systems of which certain expressions are true in the much specified classes of the structures of time.

---

138 About mereology cf. *Metamereologia* (Metamereology), Andrzej Pietruszczak, [217, 218].

Part II The issues

5 The problem

5.1 Aristotle and *The Sea Battle Tomorrow*

The ancient philosophers combined the issues of modality, time and determinism. Undoubtedly, Aristotle was one of the first to address these issues in his writings, and his reflections entered the canon of philosophy and logic under a keyword: *sea battle tomorrow*. At this point, however, the certainty ends. For the standpoint Aristotle presents in *On Interpretation, IX* is quite unclear and provides many doubts for the reader as concerns interpreting the intentions contained therein. In his book, *Time and Necessity. Studies in Aristotle's Theory of Modality*, Jaakko Hintikka states that it is the basic difficulty in understanding the idea of Aristotle:

> The most important problem raised by Aristotle's discussion of singular statements concerning the future in De Interpretatione 9 is the question: what is the discussion all about? Scholars disagree not only about the details of Aristotle's discussion; they have given different answers to the question: what is Aristotle's problem in De Int. 9? What is the view he wants to refute there, and what is the view for which he wants to argue? ([104], p. 147).

The issue raised by Hintikka is probably one of the fundamental reasons of such divergence of the interpretations of Aristotle's views on the issue of logical value of the sentences about the future, and on the principle of bivalence. This is not the only difficulty. Even if we were to acknowledge that the Stagyrite discusses a problem of logical determinism and its relation to the principle of bivalence in an attempt to strike a happy medium between the range of the application of the principle of bivalence and the justified claim that a logical value of sentences about the future states of affairs is incidental, there are still many issues raising doubts[139]. Another important and also unclear issue arising

---

139 The principle of bivalence is ensured by the law of non-contradiction $\neg(A \wedge \neg A)$ which states that each sentence carries at most one logical value; and the law of excluded middle: $A \vee \neg A$ that ensures it carries at least one logical value. However, considering that it is a modern distinction (associated with a clear differentiation between the syntactic and the semantic concepts) and that Aristotle had accepted the law of non-contradiction, I shall omit this distinction while writing about the law of excluded middle and the principle of bivalence interchangeably. I would like to add that in this regard I am developing a view expressed by Jaakko Hintikka who states that he has not found such distinction within the Aristotle's writings ([104], p. 148).

when attempting to interpret *On Interpretation, IX*, is the question of Aristotelian modalities.

5.1.1 Primary problems concerning modality

The modal notions Aristotle operates with seem to near his stance to the standpoint of determinism of the future, even though he has been a proponent of indeterminism of the future (Hintikka J. [104], p. 113), which is expressed precisely in the considered issue of the tomorrow's sea battle. These concepts are present in the considerations about the truthness of sentences[140], time and determinism, for the very reason, that such modal phrases as *necessary*, *possible* or *contingent* the Stagyrite refers to and defines by the concept of time. Between thus understood *necessity* and *possibility*, there are, however, certain relations that, by negation, can be expressed in the following very modern manner:

(1) *A sentence is possible if and only if its negation is not necessary,*
(2) *A sentence is necessary if and only if its negation is not possible*[141].

The reference of modality to time has its grounds in so-called (*principle of plenitude*) of which acceptance Hintikka attributes to Aristotle. This principle states that *every possibility is realized in a certain period of time* (cf. [104], pp. 93–113)[142]. According to this principle, a possible sentence is a sentence that sometimes is true[143]. However, in relation to such approach, in the very text of Aristotle there are certain counter-examples. For instance, Aristotle writes that: *it is possible that this coat may be cut in half, and yet it may not be cut in half, but*

140 For clarity, I shall add that in ancient times there had been a prevailing concept of sentences or propositions temporally indeterminate, that is corresponding to the elements from set $\mathbf{S}_{\mathrm{TU}}$. This applies not only to Aristotle (Hintikka J. [104], p. 70), but — as shall be seen in due course — also to the Megarians. In the further part when considering the sentence 'There will be a sea battle tomorrow.' I shall, however, use a sentence in a temporally defined form 'In the time *t* there will be a sea battle,' I shall try to circumstantiate it adequately.

141 *Contingency* according to Aristotle, means as much as *not necessary, but possible.* Sometimes it accounts for controversy, as in certain fragments, Aristotle seems not to distinguish these two different concepts (cf. Łukasiewicz J. [170], pp. 207–211, Kneale W., Kneale M. [133], pp. 84–86, Hintikka J. [104], pp. 27–31).

142 Although, as Hintikka writes, Aristotle, unlike Diodorus Cronus, never defines general modal notions in purely temporal terms ([104], p. 102).

143 Whereas, according to (2), a necessary sentence is a sentence that is always true

wear out first ([11], 19a, p. 76)[144]. In the temporal interpretation of the possibility, presented above, prior wear out of a coat excludes cutting it in half, thus makes it impossible. Such point of view — as we shall see, close to the notion of Diodorus Cronus — results in certain deterministic consequences, as it excludes unrealised possibilities. However, the principle of plenitude, the origin of the definition of possibility, does not need to be applicable to individuals, but may apply either to the types of individuals or to the types of the states of affairs (Hintikka J. [104], pp. 100–101). *The possibility that the coat is cut in half* means accordingly that some coat at some time is cut in half (in the whole period of time), but not each coat. Such understanding of the definition of possibility encourages the tension between logical determinism, resulting from lack of acceptance for unrealised possibilities, and the Stagyrite's indeterministic inclinations disappear.

The distinction between the absolute and relative possibility and necessity, present in the Stagyrite's writings, poses another problem (Kneale W., Kneale M. [133], pp. 92–96), for possibility and necessity may be related to varying circumstances. As mentioned above, the fact that a coat may be cut in half ceases to be true at the moment of time when the coat does not exist. For in this situation, it is necessary that the coat is not (and was not) cut in half[145]. The absolute modalities are special cases of modalities relativised to time, as by the latter they can be defined as follows:

(3) *A sentence is possible if and only if for any t the sentence is possible in t.*
(4) *A sentence is necessary if and only if for any t the sentence is necessary in t.*

Due to the fact that the fragments analysed below seem, however, to employ modalities not relativised to time, in the remarks about *On Interpretation, IX*, we shall apply the use of absolute modalities only.

The last issue, worth to add to the list of primary problems related to modalities, concerns the question whether modalities discussed in Aristotle's writing relate to sentences (or propositions) or rather to the states of affairs (or other truthmaking objects). For some of the wordings suggest also the latter option. Without resolving this issue, we shall present two non-exclusive and complementary interpretation outlines, which to a certain extent represent both the alternatives.

---

144 All the fragments of the Aristotle 'On Interpretation' come from a publication translated by E. M. Edghill

145 Various possible approaches to temporal modalities in the works of Aristotle as well as of the Megarian and the Stoical schools, including the relativisation to time (*p is possible at t*, *p is necessary at t*), have been presented by Nicholas Rescher and Alasdair Urquart ([241], pp. 4–5).

5.1.2 Interpretations of *De Interpretatione*

The reading of *On Interpretation, IX* suggests to the reader that the Stagyrite addresses the issue of the principle of bivalence in relation to the sentences about the future states of affairs; future — because he has decided on the issue of the sentences about the past and the present states of affairs.

> In the case of that which is or which has taken place, propositions, whether positive or negative, must be true or false. ([11], 18a, p. 75).

The problem arises in the very case of the sentences about the future states of affairs:

> So that if one man affirms that an event of a given character will take place and another denies it, it is plain that the statement of the one will correspond with reality and that of the other will not. For the predicate cannot both belong and not belong to the subject at one and the same time with regard to the future. (...) It may therefore be argued that it is necessary that affirmations or denials must be either true or false. Now if this be so, nothing is or takes place fortuitously, either in the present or in the future, and there are no real alternatives; everything takes place of necessity and is fixed. For either he that affirms that it will take place or he that denies this is in correspondence with fact ([11], 18a–18b, p. 75)[146].

The assumption that the sentences about the future are subject to the principle of bivalence, naturally leads to the logical determinism, for it indicates that the considered sentences have already had a determined logical value in the past, and this suggests a certain version of metaphysical determinism:

> If a thing is white now, it was true before to say that it would be white, so that of anything that has taken place it was always true to say 'it is' or 'it will be'. But if it was always true to say that a thing is or will be, it is not possible that it should not be or not be about to be. For when a thing cannot not come to be, it is impossible that it should not come to be, and when it is impossible that it should not come to be, it must come to be. All, then, that is about to be must of necessity take place. It results from this that nothing is uncertain or fortuitous, for if it were fortuitous it would not be necessary ([11], 18b, p. 76).

146 Aristotle considers here sentences and their negations with a subject–predicate structure, which are the syntactic units of the logic of names he has developed. Of course, nothing stands in the way to generalise his views in the constative sentences in general, especially that the Peripatetics (e.g. Theophrastus) have been studying so-called hypothetical sentences, or sentences consisting of propositions, not names. It is even claimed that this way they anticipated the Stoic logic of propositions (Barnes J. [23], p. 78).

The above fragments — despite many modalities appearing there[147], which undoubtedly impede their understanding — can be interpreted as an argument for the benefit of determinism, based on the principle of bivalence. This argumentation can be presented in two ways. The first way is of a linguistic nature and refers to the relation of respective modalities and time with sentences; the other way is of a semantic nature and refers to certain relations between the states of affairs and time. We do not claim one of them to be fully corresponding to the argument originally given by the Stagyrite. Both ways convey certain philosophical perceptions hidden within the cited extracts (the fragments are: 18a–18b, [11], pp. 75–76).

The first one would be as follows. If assuming that the sentences about the future[148] are subject to the principle of bivalence, then any sentence stating about the future at the moment n is true. Assuming that the following sentence is true:

(5) *At moment n it is true that at moment t_1, later than moment n, there will be p*[149],

assuming, however, the truthness of sentence:

(6) *At moment t_1 there will not be p*,

147 Readers of Aristotle, in accordance to the previously suggested uncertainties, encounters a problem of the types of modalities. For once the Stagyrite refers modalities to the sentences or propositions, and another time to the states of affairs. His concept of modality Aristotle developed in the *Prior Analytics, III, XIII*, and *VIII–XXII* [13] and in the very *On Interpretation, XII, XIII*, and also — as can be seen — *IX*.

148 Note, that if a sentence is to state about the future, then only in reference to a certain point of present time, it should either have a structure of a sentence determined in time within a certain chronological system including a present point of time or it should be formulated in an adequate grammatical tense indicating that it states about certain future states of affairs. Then, in the both cases, the matter involves *de facto* a sentence determined in time as follows: 'At t is p.', or with a determined n: 'There is such t later than n, that p is true at t.'. (More accurate clarification for the presented differentiations will be discussed in chapter six, entirely devoted to these issues.) It means that despite previously suggested interest in the ancient sentences not determined in time, Aristotle considers a sentence determined in time. For, since Stagyrite claims that in case of *that which is or which has taken place*, then *propositions, whether positive or negative, must be true or false*, then assertions or negations should indicate a moment or an interval of time they refer to. For example sentence: 'Greece is going bankrupt.' has been false in a certain moment, but now it is true, and maybe one day it will become false. And yet sentence: 'Greece was going bankrupt.' will always be true in the future.

149 Let p symbolise e.g. a sentence 'There is a sea battle.'

we get a contradiction, as the truthness of sentence (5) entails[150], that:

(7) *At moment t_1 it will be so that p.*

But how to deduce from this fact a necessity of sentence (7)? It could have been e.g. as follows: since it is not possible for the contradictory sentences to be true, it is necessary for their contradiction to occur, that is the very law of excluded middle. Concluding that sentence (7) is necessary, from a fact that sentence (6) does not occur, would, however, require a rule that is not commonly accepted[151].

However, this conclusion can be drawn based on other modern modal logic schemes of which prototypes can be found in the ancient works, especially in the texts of Aristotle, enriched with temporal operators and with acknowledgement that sentence (7) results from sentence (5). Informally, this reasoning would be drafted as follows[152]. Since assuming that (5) $\models$ (7), then also $\models$ (5) $\rightarrow$ (7). Currently, applying Gödel theorem, axiom *K* and a classical logic, we would get that $\models \Box(5) \rightarrow \Box(7)$[153]. Nonetheless, Aristotle claimed that the sentences stating something about the future or present carry a determined logical value. Øhstrøm and Hasle believe that, according to Aristotle, necessity can be attributed to such sentences ([205], p. 11)[154]. The point here is that the true sentences about the

150 *Entailment* can be understood as a logical consequence in a certain logic taking into account temporal relations; simplifying, e.g. every time the sentence (5) is true, also the sentence (7) is true, or that the conditional sentence (5) $\rightarrow$ (7) constitutes a tautology of this logic.

151 I mean here a reasoning called *modal fallacy* of the following structure: $\Box(A \vee B), \neg B \models \Box A$.

152 This issue will be discussed more accurately in another study. I shall only outline its general framework. Symbols $\rightarrow$, $\Box$, $\Diamond$, $\models$ respectively mean material implication, necessity, possibility and entailment and logical truth in a logical system, where the presented reasoning can be formally reconstructed. This system naturally requires additional temporal functors, omitted herein. These are concealed within the considered sentences (5) and (7).

153 Martha and William Kneale claim that in the works of Aristotle there is a rule corresponding to the modern rule in a form of $\models A \rightarrow B$, then $\models \Diamond A \rightarrow \Diamond B$, attributing it also to the second premiss of **RDC** ([133], p. 92, p. 119). However, we know that by the definitions of modality, accepted by Aristotle, it is equivalent to the rule $\models A \rightarrow B$, then $\models \Box A \rightarrow \Box B$. Therefore, this proof does not require to be based on the equivalent of axiom *K*: $\Box(A \rightarrow B) \rightarrow (\Box A \rightarrow \Box B)$ and the Gödel theorem (if $\models A$, then $\models \Box A$).

154 This is also confirmed by Hintikka, who claims that Aristotle, in his writings repeatedly states that all true statements about the future and the present are necessary, and thus are subject to the rule $\Box A \vee \Box \neg A$ ([104], pp. 164–165), as well as Martha

future and the present will not change their logical values in future moments, and thus it is not possible for them to be false (cf. one of the above-cited fragments of *On Interpretation, IX*, 18a, p. 75). It establishes a principle that if the sentence states about the future or about the past, it is a necessary sentence. A sentence such as (5) is a subject of this principle, for it states about moment n. Accordingly, there is $\models (5) \rightarrow \Box(5)$, which along with a previous formula results in $\models (5) \rightarrow \Box(7)$, thus, if it is true that something will be in the future, then it is necessary that in the future it is such. This is a general conclusion for all sentences such as (5) and (7). Accepting the law of excluded middle for sentences such as (5), we can declare without exception sentences such as (7) to be necessary[155] or their negation to be necessary. The above-outlined reasoning allows to circumstantiate the following troublesome conclusion: $\Box(7) \vee \Box\neg(7)$[156]. One of the strategies to avoid these deterministic consequences is either to reject the law of excluded middle in an unlimited form or the other premisses employed in the reasoning.

Even though the ascribing of certain principles of propositional logic to Aristotle may turn out to be erroneous (for the Stoics developed the logic of propositions), we have acted in accordance with the observation of Jan Łukasiewicz that seems accurate[157]: *it can be called the Aristotelian modal propositional logic,*

---

and William Kneale ([133], p. 119), finding an equivalent for modal rule for true sentences about the future ($A \rightarrow \Box A$) in *Nicomachean Ethics*. In the above approach, I am demonstrating how to extend the principle indicated by Hintikka onto sentences about the future; which actually softens the asymmetry of time, if based on qualifying the sentences about the future as necessary, and the sentences about the past as possible.

155 My interpretation here differs significantly from e.g. interpretation of Jan Łukasiewicz, for Łukasiewicz excludes temporal characteristics of necessity and claims paradoxality of the thesis: if A, then $\Box A$, by assigning the status of necessity to the factual truths ([170], p. 207). However, if we understand the necessity like Diodorus did, i.e. that *the necessary is what is true now and always will be true*, then declaring necessity informs about a rather trivial fact, that a true sentence determined in time (which in fact is a sentence 'At t is that there will be a sea battle tomorrow') is always true in the future. Whereas the question of its logical value at a moment earlier than t is problematic.

156 According to Hintikka precisely, such thesis is one of the issues of Aristotle's discussion within the problem of *tomorrow's sea battle* ([104], pp. 163–164.)

157 The historians of logic, e.g. the mentioned above William and Martha Kneale, state that in Aristotle's writings there can be an external use of modality in relation to the whole sentences distinguished from the internal use with regard to the relationship between terms in his logic of names ([133], pp. 90–91). Herein, I have employed the external use of modality in relation to sentences, regardless of their structure.

since some of his statements are sufficiently general to include all types of sentences [and not only the categorical statements], *and some of the sentences he formulates himself using sentential variables* ([170], p. 181, author's translation). According to Łukasiewicz, in the Stagyrite's works, there can be found all modal premisses we have employed while concluding about the future sea battle: axiom K is present in Aristotle's writings *implicite* based on the principle, that if sentence A results in sentence B, then if $\Box A$, then $\Box B$, that is similar to the monotonicity rule for a necessity operator (pp. 186–187, cf. Kneale W., Kneale M. [133], pp. 86–87); there is also present an equivalent of axiom T: $\Box A \rightarrow A$ and the hypothetical syllogism. An equivalent of the Gödel theorem can be in turn considered a symbol of necessity before syllogisms, or schemes of correct reasoning regardless of the content (p. 195)[158]. An explicit attributing these formulas to Aristotle would have been rather risky. It is though understandable to notice certain pre-formal prototypes in his writings. Apart from their occurrence, there is a temporal rule missing in the Stagyrite's works that would justify a transition from the sentences like (5) to the type of sentences such as (7). In order to formulate such rule, there should be some relevant temporal operators added to the language. Within logic that would have included the said modal and temporal principles, the thesis would have been the above-mentioned formula: $\Box(7) \vee \Box\neg(7)$.

Aristotle's argument could have been, though, interpreted in a more philosophical and semantic manner, without the use of a particular logic, and the question of temporal rule and modality could have been referred to a succession of states of affairs. Let us consider the following sentence:

(8) *There will be a sea battle tomorrow.*

Today, we do not know if this sentence is true. However, accepting the principle of bivalence for such sentences, we know that it carries one of the two logical values. Since it is a simple sentence, according to the definition of truthness and falseness presented in the first chapter, if the sentence is true, then tomorrow there will be located a relevant verifier, if it is false, the verifier shall not be there.

---

158 Łukasiewicz writes that *a symbol of necessity placed before a syllogism proves that it is not a conclusion that is necessary, but a connection between premisses and a conclusion* (author's tranlsation). Aside from these rules, in Aristotle's works, Łukasiewicz indicates an occurrence of a consequence of axiom $T: A \rightarrow \Diamond A$ ([170], p. 183) and many other axioms, statements or principles of modern sentential modal calculus (cf. pp. 180–211). Łukasiewicz also describes a certain modal logic system, which would meet the needs of Aristotle [171]. The presented prototypes of the principles of modal logic by all means allow various alternative proofs of thesis $\Box A \vee \Box\neg A$.

In any case, this issue is preordained today, and because any simple sentence or its negation can be handled in such manner, this approach leads to determinism in both logical and metaphysical meanings in relation to all these states of affairs that can be expressed with the use of such understood simple sentences. Even if a sentence of the type (8) is not necessary in a sense of *de dicto*, then in time determined in it, there will occur the described state of affairs, and, at least this way, it is necessary and non-coincidental — for they are not subject to any additional factors that would have changed that in the meantime. Aristotle himself states that:

> Wherefore, if through all time the nature of things was so constituted that a prediction about an event was true, then through all time it was necessary that that should find fulfilment; and with regard to all events, circumstances have always been such that their occurrence is a matter of necessity. For that of which someone has said truly that it will be, cannot fail to take place; and of that which takes place, it was always true to say that it would be ([11], 19a, p. 76).

Regardless of the interpretation of necessity, if we consider it either an essential element of a sentential modal system or an irrelevant element of persuasion and rhetoric, the logical value of the sentences about the future compromises casualness, what contradicts Aristotle's intuitions:

> (…) it is not of necessity that everything is or takes place; some facts come to be by chance (…)[159] ([11], 19a, p. 77).

But since Aristotle rejects such deterministic consequences, he must also reject the premisses they result from. Whereas the principle of bivalence is one of them:

> Everything must either be or not be, whether in the present or in the future, but it is not always possible to distinguish and state determinately which of these alternatives must necessarily come about. Let me illustrate. A sea battle must either take place tomorrow or not, but it is not necessary that it should take place tomorrow, neither is it necessary that it should not take place, yet it is necessary that it either should or should not take place tomorrow. ([11], 19a, p. 77).

According to this fragment, Aristotle does not reject the principle of bivalence, he only modifies it. The alternative of two contradictory sentences about the future is true, but none of these sentences has to be true[160].

---

159 In this fragment, Stagyrite emphasises, among others, the issue of human causativity that can modify future events (cf. Faye J., Scheffler U., Urchs M. [72], pp. 21–22).

160 Therefore, in a realm of sentences about the future, it is not about a regular extensional alternative. For this one, if it is true, requires at least one of its two parts to be true. The alternative Aristotle seems to be suggesting is true, even though none of its two

The interpretation of the above excerpt of the Aristotle's work is not though so obvious as we have stated. For in literature, there are different views than we have offered, which, apart from not being compatible with our proposition, are also mutually exclusive. Jan Łukasiewicz believed that Aristotle advocated for introducing a third logical value for the sentences about the future([170], p. 209) (the fact, that Aristotle rejected classic logical values for sentences about the future had been emphasised in the Middle Ages as well (cf. Øhrstrøm P., Hasle P. F. V. [205], p. 12)). This view has been also shared by Prior A. N. ([230], p. 86). On the other hand, in the vein of the principle of bivalence, N. Rescher defended the interpretation of these fragments [242]. Another one, a newer interpretation has been proposed by H. B. Andersen and J. Faye (Øhrstrøm P., Hasle P. F. V. [205], pp. 13–14).

The above discussion we shall summarise as follows. Aristotle certainly claimed that the sentences about the past and the present have logical values and the principle of bivalence can be applied to them without any reservations. If they carry a logical value in a moment of time, then it does not change and, according to the above-mentioned reflection of Øhrstrøm and Hasle, we can assign them necessity. Naturally, this is about necessity related to the nature of time and non-reversibility of history (this belief, as already mentioned, will be also shared by Diodorus Cronus). Therefore these authors call Aristotle a *past and present determinist* ([205], p. 11). Aristotle, however, undoubtedly rejected the belief that sentences about the future were also necessary. Seeing the source of this believe in the principle of bivalence in relation to this type of sentences, he rejected it or at least modified it. However, it is certain that for remaining types of sentences this principle has been fully accepted by him.

5.1.3 *On Interpretation, IX vs.* The Reasoning of Diodorus Cronus

A reflection on Aristotle's views raises a question about the relationship between his considerations and **RDC**. As for historical relations and possibilities of Aristotle's direct influence on Diodorus, opinions vary[161]. Obviously, there is a substantive relation between both issues, which can indicate that they constitute

parts is true. Only in the future one of them shall become true. It opens an interesting opportunity for the studies on the formal characteristics of such *indeterministic* alternative.

161 Some researchers, e.g. K. V. Fritz, claimed that both issues are related, others, e.g. D. Frede, believed otherwise (cf. Döring K. [64], p. 133). **RDC** has been regarded an attempt of polemics with Aristotle's considerations also by Prior ([231], p. 138).

a part of a wider discussion carried in ancient times, which will be partly presented in the next subchapter. Participation in this wider debate would suggest though a possibility that Diodorus did polemise with Aristotle.

However, regardless of the historical facts, **RDC** can be considered a polemics with the views of Aristotle. The following arguments advocate this idea.

Both philosophers discuss relations between logical values, time and modalities, and it locates them in the same subject area. While Aristotle arguments in favour of indeterminism, Diodorus' reasoning has been considered in ancient times as an argument in favour of fatalism |(cf. Döring K. [64], p. 134). Both argumentations are also related by accepting what is called past and present determinism, and what in **RDC** constitutes one of fundamental evidential premisses.

A crowning argument in favour of the postulated relation would have been however a constatation of the following fact. In the end, from the cited excerpts, Aristotle claims that neither a sentence about the future nor its negation are necessary. Hence, it results in a rule that for any sentence about the future, possible is alike the sentence and its negation, what is expressed in the view of indeterminism[162]. Whereas Diodorus claims that a sentence is possible if it is true or will be true. A sentence that neither is nor will be true, according to this definition, is not possible. However, it is known that of the two contradictory sentences only one will be true. Therefore, according to Diodorus' definition, only one is possible, and it contradicts the principle of Aristotle. However, this observation is not so clear if related to sentences from set $\mathbf{S}_{\mathrm{TU}}$, for then, even in accordance with the definition of Diodorus, both sentences can be possible.

Some historians openly balance the reasoning of Diodorus against the views of Aristotle, implying that the former is set against the latter[163].

---

162 An order of quantifiers is an important issue in this argument — however it is difficult to identify unambiguously. If assumed that all remains within a range of a large quantifier, then it means that the subjects of modality are only empirical, synthetical sentences. For this principle would have been non-intuitive if considered sentence has been counter-analytic, purportedly stating about the future — purportedly, for it would be false regardless of empiricism.

163 I mean e.g. Giovanni Reale, who emphasises, that Diodorus is defending the Parmenidean argument about the appearance of a change, and to this end he attacks Aristotelian differentiation into potentiality and actuality, attempting to prove that there are no potentialities that are not actualities, therefore, that there is no change (cf. Reale G. [236], p. 87–91). Remarks similar to the ones of Reale also come from Martha and William Kneale, claiming that **RDC** is directed against the views expressed by Aristotle in *On Interpretation, IX* ([133], p. 122). They also put in that the Megarians

The further studies on **RDC** will be based on a part of the principles accepted by Aristotle. Even about the question of the principle of bivalence — considering its deterministic consequences — we shall agree with Aristotle. The strategy of the formal research on **RDC** shall be described in the further subchapters about the very sentences about the future. Whereas currently we shall discuss the historical evidence concerning Diodorus' views and his reasoning. On the basis of them, we shall determine the general postulates that will be taken under consideration in the discussion about different reconstructions of **RDC**.

5.2 The reasoning of Diodorus Cronus

Diodorus Cronus had been born presumably around 350 before Christ, and a part of his life he spent at the court of Alexandria in the reign of Ptolemy I Soter (Mates B. [178], p. 234). It is known that he was studying in the Megarian school of which founder had been one of the students of Socrates — Euclid. In turn, the students of Euclid were such known philosophers as Thrasymachus (whose successor, through Stilpo, was Zeno and the Stoic school), Ichthyas, who indirectly, through Apollonius Cronus, could have had some influence on Diodorus; and Eubulides, the creator of the liar antinomy (Mates B. [179], pp. 12–13). Since Stoic logic, constituting anticipation of modern propositional logic (cf. [200], Nasieniewski M.), had been under the influence of logic reflections of the Megarian school, then considering also the paradoxes that preserved from the Megarians, one should not be surprised by the fact that the Megarian philosophers had been considered experts in dialectics. Diodorus Cronus apparently stood out in this art so much that the contemporaneous people named him the *Logician* or *The most logical.* His alias *Logician* has even been inherited by Diodorus' five daughters who have also dabbled in logic([179], p. 13). Diodorus' passionateness and commitment in the theoretical reasoning is shown in a story of Diogenes La'rtius, where, during one of the feasts at the court of Ptolemy Stilpon, he had presented a logical problem that Diodorus could not solve on the spot. Only later he presented his solution in writing, nevertheless considering the incident his failure. The Megarian philosopher was to be greatly concerned about it, that from then on he had been depressed, and before long he even died of despair (Diogenes Laertios [60], II 10, p. 139).

(thus Diodorus as well, as he belonged to an early generation of the Megarian school) have been rejecting Aristotle's differentiation into actuality and potentiality, trying to find new concepts of modality instead (p. 117). This is apparently the closest approach to the one I shall develop. I will though omit the historical dispute, and I shall develop the reasoning of Diodorus in a purely logical manner, without entangling in it the rules of contemporaneous metaphysics and physics.

Moreover, as Mates writes, *at that time, logical problems were considered more serious than nowadays* (Mates B. [178], p. 234). Because, as mentioned many times already, philosophy meant for Ancients more than just a trade. Philosophy was a way of life[164].

It is being said that the name Cronus, that is an old fool[165], Diodorus could have gotten the name precisely for his indolence during the feast. However, considering his earlier-mentioned aliases, it seems more probable that — as other records show — he was named Cronus after his teacher Apollonius Cronus (Mates B. [179], pp. 12–13).

Not much is known about Diodorus' philosophy. Historical materials available today come solely from doxography, quite often unfavourable towards the Megarians[166]. There are, however, reliable testimonies for Diodorus formulation of some concepts crucial for logical reasoning, in particular concerning issues of time and modality[167].

One of the concepts defined by Diodorus describes the issue of the truthness of conditional propositions, or, as we would say today in a more nominalistic manner, the truthness of conditional sentences. In Diodorus' times, this problem had been so widely and deeply discussed that a certain Kallimach apparently noticed, that *even the crows on the roofs caw about which definition of the truthness of conditional sentences is correct* (Mates B. [178], p. 234). Other concepts the Megarian considered were directly associated with the subject matter of this study. Diodorus has given the definitions of possibility and necessity on the basis of the very relation of the question of truthness of propositions to the

164 In this case, I identify the ancient philosophy with the ancient logic, since even though the Peripatetics considered logic a tool (it was not any of the three types of knowledge described by Aristotle), the philosophers related to the Stoic school considered logic a part of philosophy (cf. Barnes J. [22], pp. 67–69).

165 Greek word *κρνς* refers to at least three things: mythical Cronus, planet Saturn and as a generic name may mean as much as an old fool (Liddell H. G., Scott R. [159], p. 998).

166 Probably most of the preserved ancient fragments stating about Diodorus and so-called *his circle*, are collected in the work of Klaus Döring *Die Megariker. Kommentierte Sammlung der Testimonien* [64], numbered 96–143. In reference to the fragments of our interest from doxography, I will first give their number after Döring, not their own adopted pagination. Whereas, while citing I shall refer to their English translation.

167 Diodorus also dealt with other, *stricte* philosophical issues, but he practised philosophy using dialectical methods (today it would be called logical methods), as evidenced by e.g. four arguments against the existence of motion, ascribed to him (Sedley D. [254], pp. 356–362).

time parameter. Whereas trying to convince his opponents, that his definitions are correct, he apparently conducted his famous reasoning.

5.2.1 The definitions of modality *vs.* Diodorus' reasoning

Possibility, necessity and other modalities are not independent (e.g. syntactic), but refer to something. This *something* are e.g. propositions or sentences, as well as states of affairs, events or predicates. It seems that the reasoning of Diodorus Cronus is about modalities in relation to propositions, although we shall also pay attention to other opinions, which claim other possibilities[168].

However, remaining on the propositions in the context of Diodorus' modalities, more should be said about them, for the term *proposition* is polysemic. Unfortunately, we know nothing about Diodorus' approach to the issue of propositions. Since his considerations remain related to the views of the Stoic school, which had imported the interest in logic from the Megarians, and Diodorus' views had been discussed by its scholars; nothing else can be done but to take the opinion, that the semantic views are in essential points consistent with the thoughts of Diodorus, at a face value; what also does cited Mates in his book.

According to Mates, the Stoics' approach to signs, meaning and range had been very similar to the modern distinctions presented by Frege, who has been mentioned in the first chapter (cf. Mates B. [179], p. 34 and others). The Stoics defined proposition as a full *lekton*, self-predicative (p. 43), understanding the term *lekton* more of less in the same way as the word *Sinn* is understood in Frege's terminology[169]. The examples of propositions are e.g. *Hannibal was Carthaginian, Socrates is bold.* Accordingly, for the Stoics, proposition has not been a meaning of individual sign that builds a sentence, but a meaning of the

168 Jonathan Barnes believes that the discussion about four modalities: possibility, impossibility, necessity, and non-necessity has been intended to demonstrate which definitions are adequate in a meaning of correctly describing the world. The considered modalities would have in fact also been of metaphysical nature. Furthermore, the problem of relationship between falseness of the sentences about the future and their presumptive impossibility would have perhaps added these discussions a practical dimension, moving them towards the issues of determinism and free will (cf. [23], p. 86).

169 It is worth to add, however, that Hintikka criticises such interpretation of *lekton*, writing that it relates more to sentences or propositions temporally indeterminated ([104], p. 71). In this matter, he refers to the opinion of the Kneales. It does not change the fact that **RDC** leaves a wiggle room for both types of sentences or propositions.

very sentence. This school divided propositions in terms of the two major criteria: syntactic and semantic.

In terms of the syntax, propositions were divided into atomic and molecular. Atomic propositions represent what we would call today simple propositions — their parts are not propositions. These are structured of a subject and a predicate, and there are no logical connectives involved. Whereas molecular statements are simply statements which are not atomic. Amongst connectives, the Stoics used counterparts of today's sentential negation, conjunction, disjunctive and conjunctive alternative, and conditional period, variously interpreted. Additionally, they employed different non-extensional functors, used to create causal propositions or propositions stating about probability.

Due to an issue of logical value, propositions were divided into true and false, whereby each proposition was to be either true or false. This approach has been intentionally inconsistent with Aristotle's thinking. The Stoics were supposed not to deviate from this rule towards propositions about the future and, according to Boethius, it was to aim intentionally into the Stagyrite's views([179], p. 45). The fiercest defender of the law *tertium non datur*, according to Cicero's *De Fato*, had been Chrysippus. This law have been formulated based on the assertion that *each proposition is true or false* and that *the alternative for a proposition and its negation is by necessity true*, by alternative meaning disjunctive alternative, for the Stoics preferred such in the proposed five basic types of arguments (p. 46)[170].

Currently, we shall continue to the issue of possibility and necessity. The dispute on their substantially accurate definition had been a topic of considerations primarily of Diodorus, Philo and Chrysippus.

The definition of possibility of Diodorus could have been expressed in the following statement: *what is possible is what is or will be the case* — referred by Alexander of Aphrodisias (Mates B. [179], p. 57, cf. Döring K. [64], frag. 135, p. 41)). Boethius has given definitions of all other relative concepts:

i) impossible is what being (now) false, will not be true,
ii) necessary is what being (now) true, will not be false,
iii) non-necessary is what is false or will be false (Mates B. [179], p. 57, por. Döring K. [64], frag. 140, pp. 42–43).

Using the same style as in the definition of possibility given by Alexander, we would define as follows:

170 It is worth to remind that, according to the ancient evidence, the Stoics were the proponents of the strict physical determinism and as one of the first introduced doctrine into their cosmological considerations (cf. Furley D. [78], pp. 450–451).

iv) possible is what is true or will be true.

These four definitions lead to a conclusion that Diodorus viewed possibility and necessity as the concepts that combined with negation are interdefinable. For, according to the above definitions, we can determine in a neutral style that:

v) something is necessary if and only if it is not possible that it is not,
vi) something is possible if and only if it is not necessary that it is not.

In the historical evidence of Alexander, we find such description:

> According to him [Diodorus] it is possible that I am in Corinth if I am in Corinth or if I plan to be in Corinth; but if I was never to be in Corinth, it would have not been possible. Similarly, it is possible that a child will become a grammarian if one day the child will become a grammarian. And in order to determine this concept, Diodorus propounded his "master" argument. (Mates B. [179], pp. 57–58, cf. Döring K. [64], fragm. 135, p. 41)).

As shall be presented below, it is reasonable to suppose that Diodorus based on the concept of possibility. It means that, while reasoning that the above definition of possibility is relevant, he could subsequently try to derive from it the remaining concepts. Apparently, the purpose of Diodorus' reasoning has been to stabilise a certain definition, not the opposite, that is formulation of the reasoning on the basis of the definitions of modality. Essentially, so-called Diodorus' definitions of modality are not, therefore, definitions but theorems based on a certain set of premisses, by means of which master reasonings can be carried out.

According to the opinion of the Ancients, the reasoning of Diodorus[171] emerged from a certain trilemma. The assumption of truthness of the three premisses was to lead to a contradiction. As the two first premisses passed for convincing, Diodorus rejected the third one, obtaining a definition (or a part of it) of the above-mentioned concept of possibility. Although the information about the process of the reasoning is scarce, the premisses we know from Epictetus's relation, which in an English translation, can be found in a chapter of a distinctive title *Against those who embrace philosophical opinions only in words.*.

> The argument called the ruling argument appears to have been proposed from such principles as these: there is in fact a common contradiction between one another in these three propositions, each two being in contradiction to the third.

---

171 The reasoning of Diodorus have been called *ὁ κυριευον*, what translates as masterful (Mates B. [179], p. 58) or, as will be seen in due course, ruling, masterly. All these phrases are to suggest its irresistible force of influence.

1. that everything past must of necessity be true[172];
2. that an impossibility does not follow a possibility;
3. and that a thing is possible which neither is nor will be true.

> Diodorus observing this contradiction employed the probative force of the first two for the demonstration of this proposition: That nothing is possible which is not true and never will be (Epictetus [69], II 19, pp. 173–174, cf. Döring K. [64], frag. 131, pp. 214–215).

The contradiction between these three propositions was, as it seems, universally accepted by contemporary philosophers involved in the discussion. Mates claims that nobody questioned the fact that they were inconsistent ([179], p. 59). Various philosophers, however, emphasised different sources of these inconsistencies, in fact realising all possibilities to reject one of these propositions. Again, Epictetus's evidence comes to aid:

> Now another will hold these two: That something is possible which is neither true nor ever will be: and That an impossibility does not follow a possibility. But he will not allow that everything which is past is necessarily true, as the followers of Cleanthes seem to think, and Antipater copiously defended them. But others maintain the other two propositions: That a thing is possible which is neither true nor will be true: and That everything which is past is necessarilly true; but then they will maintain that an impossibility can follow a possibility. But it is impossible to maintain these three propositions, because of their common contradiction.[...] but I have received this story, that Diodorus maintained one opinion, the followers of Panthoides, I think, and Cleanthes maintained another opinion, and those of Chrysippus a third (Epictetus [69], II 19, p. 174, por. Döring K. [64], frag. 131, p. 215).

Thus, Epictetus claimed that the three above propositions are contradictory. If trying to indicate the reason for these contradictions, an attempt to analyse them poses a very fundamental problem we have omitted for a moment: what actually are these three sentences about? Let us see e.g. the first sentence from Epictetus's account; it states that: *what is past and real is necessary*. English translation offers a word *real*, but other translations leave no doubt that the term in question is a modern word *true*. In English translation quoted by Nicholas Rescher [243] and Frederick Seymour Michael [190], this sentence stands: *what is past and true is necessary*, suggesting that it is about something that is past and true, and therefore necessary. As Michael accurately noticed: only propositions (or sentences) can be true, and events and states of affairs can be past, not conversely ([190], p. 229).

---

172 It would be about the type of necessity called by scholastics the necessity *per accidens*: the past states of affairs not only belong to the history of the world, but also are necessary, because they cannot be changed in any way (cf. Faye J. [71], s. 104).

This differentiation allows two ways of interpretation regarding the subject of Diodorus' reasoning. Obviously, it is not about historical, but material interpretations[173].

For on the one hand it is possible to formulate this premiss so it states about events, states of affairs etc., simply about the objective matter of the language — about the world. Then it would take the following form: *an event that occurred in the past is necessary* with additional consequences for the remaining premisses. It would have been an interpretation involved in the problem of the ontology of the world, time and modalities, not of the propositions[174].

However, according to the prior remarks, Diodorus rather attempted to substantiate the relevant modal concepts in regard to propositions. Historically correct interpretation should have thus involved the reasoning into the ontology of truth, time and language, interacting with the world only secondarily. Therefore, there is only second option left. The first premiss states about propositions, it declares that true propositions about the future are necessary. This is the way we shall explore, not only because it seems historically more accurate, but also due to its certain stabilisation in the philosophical and logical literature — as substantially accurate — which shall be reflected in the presented reconstructions of Diodorus' reasoning. The change we shall introduce, relates only to the fact that we will not be considering propositions, but sentences we have assigned to him in the first part of this study. Prior to formulation of the three sentences that constitute the premisses of the ruling reasoning, in the styling we

---

173 For the opinion that **RDC** regards a linguistic, not an objective aspect, is historically established. At this, however, an agreement ends. Despite the above-described Mates's stance, Martha and William Kneale write that we do not know what Diodorus' modalities refer to, whether these are: *an indicative sentence, a proposition in the modern sense, or some third thing distinct from either of these* ([133], p. 118). They state that the whole reasoning is based on a certain uncertainty related to confusion between constative sentences in different grammatical tenses and propositions they express. (pp. 121–122). This is of course an acceptable alternative, that, from historical point of view, perhaps even occurred. The modern attempts of reconstruction, however, allow to prescind from it by outlining the theory of **RDC** in a way that is greatly independent from the facts unknown to us. Such partly ahistoric point of view also in my study is assumed.

174 Klaus Döring indicates a problem with the clear determining the modalities in question. A Greek word $\alpha\lambda\eta\vartheta\varepsilon\varsigma$ may though mean both *true*, as well as *real* ([64], p. 135).

have adopted, two more matters require attention. For even within the accepted interpretation there are certain controversies present.

One of them relates to the second premiss, which, in the new stylisation, would have stated as follows: *an impossible sentence does not result from a possible sentence*. However, it may mean at least two things (cf. Øhrstrøm P., Hasle P. F. V. [205], p. 16). In the first instance, suspicion arises that it is about a logical consequence, that is, calling spade a spade, a logical consequence in a given logic in which Diodorus' reasoning is being carried out[175]. The second option accepted in the literature is the interpretation of this premiss as a negation of a certain time sequence, not logic one[176]. For clarity we shall illustrate as follows. Let us assume, quite sophistically, that at a certain moment t_1 it is possible that i) *equation* $x = \sqrt{3}$ *has a solution in an integral domain*. Some time later, at a moment t_2, we assert that it is not possible that i) *equation* $x = \sqrt{3}$ *has a solution in an integral domain*. It means that qualification of something as possible changes in time. The second premiss of Diodorus' reasoning may, however, forbid it. It is not that i) is possible at t_1, and i) is impossible at t_2. Since obviously i) is not possible at t_2 (I shall add that by the virtue of relevant definitions it is not so at any moment!), it follows that t_1 is not possible. Generalising this example, it can be assumed that if a sentence is possible at a certain moment, it is possible at each subsequent one. Yet, if it is not possible at any later moment, then it was not possible in the earlier one. However such concept of possibility seems doubtful — for in an everyday life possibilities of some sentences seem to change all the time — it constitutes one of the interpretations recommended in the literature. And although this interpretation, from a historical point of view, is less acceptable according to some authors, and, according to others, more acceptable (Mates B. [179], s. 59–60), it still can be considered a certain path of philosophical and logical research.

Now, in compliance with the described difficulties, we shall formulate the premisses of the master argument:

D1 *Each sentence stating about the past is necessary.*

---

175 According to Kneales, this premiss does not pose questions and relates only to one of the principles of Aristotle ([133], p. 119), I have defined as the following rule:$\models A \rightarrow B$, then $\models \Diamond A \rightarrow \Diamond B$ (cf. also Barnes J. [23], p. 90). On the other hand, Hankinson *explicite* claims that the first and the second premisses in the Diodorus' reasoning correspond to the principles accepted by the Stagyrite in the *On Interpretation, XI* ([97], pp. 516–517).

176 Eduard Zeller is a seminal figure of such approach ([304], pp. 269–270, [305], pp. 151–159).

D2 *No impossible sentence results from possible sentence.*
D2′ *No possible sentence becomes impossible sentence.*
D3 *There exist a sentence that is possible, but neither now is true nor in the future will be true.*

The above sentences form a set of the premisses that will function as an information base. On its grounds we shall build further discussion about its formulation, not seeking an aid within historical sources any more.

Another important issue raised in literature — after determining what actually Diodorus' premisses state about — is a question about the understanding of the word *sentence* occurring there. We know that set of sentences **S** is union of two disjoint sets: $\mathbf{S}_{\mathrm{TD}}$, a set of sentences determined in time, and $\mathbf{S}_{\mathrm{TD}}$, a set of temporally indeterminate sentences. Which type thus is the type in question?

Mates points to the fact that Diodorus employed a concept of propositions that would correspond to the sentences belonging to set $\mathbf{S}_{\mathrm{TU}}$. Diodorus' propositions would not have been actual propositions in our understanding, but rather certain propositional functions with a time variable, e.g. *Toruń has 30 thousand citizens at moment t*, which change their logical values according to the value of variable *t*, and with each different value they express a different proposition. Such approach to sentences — as noticed by Mates — poses however certain problems. For the first premiss seems to state that a sentence true in the past is a necessary sentence([179], pp. 60–61). If the purpose here is that a sentence shall not change its logical value, then in the case of sentences temporally indeterminate that, after all, can change their values from context to context, the intention is wrong, unless there will be an additional assumption concerning necessity, e.g. its reference to time or to the whole context[177]. However, these detailed problems shall be dealt with in due course. For the time being, we accentuate that the attempts

177 Martha and William Kneale write that the important issue here is the difference between a sentence expressed in a past tense and a sentence referring to the past ([133], pp. 119–120). For example, true sentence 'There was a sea battle on the Baltic Sea.' is, according to Diodorus' definition, necessary, but a sentence 'There is a sea battle on the Baltic Sea.', regardless to its past truthness, is not necessary. This path is followed by a number of interpretations of the reasoning of Diodorus Cronus. Barnes adds, however, that the presented case would rather affect positive sentences relating to certain states of affairs, e.g. 'I was in Cracow' is in this sense necessary, while sentence 'I was not in Cracow' is not necessary, because it may change its logical value, if after some time I will go to Cracow ([23], p. 90). And perhaps it is not true, for sentence 'It was that I was not in Cracow' is also a sentence that does not refer to any state of affairs, but, according to the above interpretation of the first premiss, it is necessary.

to reconstruct the reasoning may be based on sentences determined in time as well as on those temporally indeterminate.

As for the issue of modal concepts defined by Diodorus Cronus, his opponents, as mentioned above, have not been sharing his views. His student Philo defined the possibility of a proposition in a quite similar manner to the modern logic approach to possibility; as a proposition, which *internal nature allows for it to be true* whereas the fact of necessity of proposition meant that *being true, by its nature it does not allow falseness* (Mates B. [179], p. 61). Philo did not agree with Diodorus' definitions of modality, claiming that a piece of wood lying on the seabed should be considered combustive, even if it would never burn (p. 14). Chrysippus's concept of modality has been quite similar to the concept of Philo (pp. 61–62). What is interesting, Chrysippus claimed in opposition to Diodorus, that the events that will never occur are still possible. Cicero mentions about it in the same work, where he writes that Chrysippus defended the principle of bivalence for all the sentences, thus, in fact, a stance, on which logical determinism is based (cf. Döring K. [64], frag. 132A, 132B, pp. 39–40)[178]. It proves that the modal concepts coined by Philo and Chrysippus could have been more general than the concepts defended by Diodorus.

Diodorus undoubtedly referred various concepts to the time parameter. There are similar accounts on this subject, in terms of the dispute about the conditional sentences which are so frequently present in every reasoning.

5.2.2 Diodorus' conditional sentences

Conditional sentence is usually understood as a compound sentence of a phrase 'if…, then…' (or other phrases of similar functions), where instead of dots there are two simple sentences. Simple, as the language rules allow building sentences in a finite manner.

The Megarians and the Stoics disputed about the conditions required for a true conditional sentence. An accurate summary of the dispute constitutes the report of Sextus Empiricus, where he sets in an order four stances, from the weakest one to the strongest one, illustrating them with examples (Mates B. [179], pp. 72–73, cf. Döring K. [64], frag. 141, 142,

---

This problem basically relates to the relation between the grammatical past tense and the negation.

178 For the stoical determinism has not been based on the concepts associated with temporal relations, but rather has accentuated the role of fate, destiny, thus certain ontological factors (Hankinson R. J. [97], pp. 526–528).

pp. 43–44). The third stance apparently belongs to Chrysippus and imposes, on the truthness of conditional sentence, similar requirements to the ones of a modern strict implication. The fourth opinion is considered somewhat enigmatic, thus we shall omit it (cf. Mates [179], pp. 64–78).

The least accurate definition comes from Philo and is basically identical to the definition of modern material implication (Mates B. [178], p. 234). Diodorus Cronus is the exponent of the second stance. This attitude states that a conditional sentence is true, *if it is not and never was possible for it to have a true antecedent and a false subsequent.* Mates, understanding this collocation as possibility defined by Diodorus, reaches some interesting conclusions. Let us remind that, according to Diodorus, a sentence is possible if it is true or will be true. Therefore a sentence is not possible if it is false and will not be true. Conditional sentence is accordingly true, if it is not and will not be that it has a true antecedent and a false subsequent. The definition of the truthness of a conditional sentence also states that it was never possible that an antecedent was true and subsequent was false, and it means that it never was that an antecedent was true and a subsequent was false, and also that it was that it will never be so. It will not be, that is an antecedent will not be true and a subsequent will not be false, neither now nor in the future. In Diodorus' definition, the first condition results from the second one by virtue of the definition of possibility. Mates believes that it has its stylistical substantiation and, following this line of thought, he interprets Diodorus' conditional sentence as a version of a material implication, true at all moments in time[179]. More generally speaking, a sentence of a form 'if *A*, then *B*' according to Diodorus' definition, is to be true if and only if for any moment of time *t*, the following material implication is true:

$$(A \text{ is true at } t) \rightarrow (B \text{ is true at } t).$$

If this interpretation is correct, then Diodorus subjected truthness of a conditional sentence to the time parameter. His approach to other extensional logical connectives is not known, but we know, however, that the Stoics school used them in a quite similar manner as we do today in the classical logic.

179 Other historians also agree with this interpretation. Barnes writes e.g. *we do not know whether, in his concept of conditional sentences, Diodorus involved some general and intuitive concept of possibility, or just his own temporal possibility. However, when assuming that the latter has occurred, then Diodorus' conditional sentence is the conditional sentence of Philo, true at every moment of time* ([23], p. 85).

5.2.3 Conclusions and indications for a reconstruction

We shall present now some postulates and remarks that will be binding in the further part of the book and which will not be again *explicite* formulated, even though the further works will have been adhering to them.

Let us begin with the question of the very reasoning of Diodorus. Jonathan Barnes believes that the reconstruction of the Reasoning of Diodorus Cronus should comply with three conditions: i) employ strictly logical measures and concepts available in ancient times, which consider the time parameter; ii) be expressible in a natural language; iii) not to be of too much complexity, so that it is possible to present it at a social meeting, *where people, during dinner, enjoy a discussion about the Master Argument* ([23], p. 89).

These requirements will have been realised to a varying degree. The presented reconstructions shall undoubtedly fulfil the two last conditions, but not the first one. This is not happening due to any purposeful reaching beyond the arsenal of logical techniques available to the Ancients. Quite the opposite. We shall try to follow the presented historical evidence, at the same time however, not excluding less accurate, but interesting, reconstructions. The precise formulating of the reconstruction requires however to employ logical means that had not been available to the ancient logicians. opposite' is a fragment. Please check if we can change this to 'It is quite the opposite.'

The name of the Reasoning of Diodorus Cronus should actually refer to such a sequence of logical operations presented by Diodorus Cronus in order to demonstrate the inconsistency of the three discussed sentences. The further consequence, that is rejection of the third sentence and acceptance of its negation, is a simple consequence of the fact that the first two were considered by Diodorus the axioms of his theory. The reasoning alone, however, no doubt about it, unfortunately had not remained. Only the three premisses and a conclusion — the definition of possibility, have survived. The crucial task is therefore generally an attempt to reconstruct this reasoning. Its reconstruction would have to be based on indicating such a non-contradictory theory T, that when attaching the three premisses $D1$–$D3$ ($D2$ naturally only in a single version)), we get a logical contradiction. By obtaining a logical contradiction, we mean a fact that by employing classical logic to the theory and the premisses we get a contradiction.

$$T \cup \{D1 - D3\} \vdash A \wedge \neg A$$

By theory T we mean certain logic L_t in compliance with the time parameter, that is superstructured over the classical logic; and a set of additional specific rules $\mathbf{R_t}$, not necessarily nonempty. Set $\mathbf{R_t}$ can contain the following rules (expressed in a given case in a language of the considered theory) that can be assigned to Diodorus and his interlocutors:

(R_t1) *If a sentence is necessary it is true.*
(R_t2) *Necessity and possibility are interdefinable by means of negation.*[180]
(R_t3) *If a sentence is now true or will be true in the future it is possible.*[181]

The language of theory T should allow to express Diodorus' premisses, and at least their temporal character, the other modalities may be left not interpreted.

Due to the fact that Diodorus' reasoning has been based on the idea of indirect proof, to capture its nature, it should be carried out in the following manner: To accept the premisses $D1$, $D2/D2'$, add premiss $D3$ and classically prove a contradiction:

$$T \cup \{D1, D2/D2'\} \cup \{D3\} \vdash A \wedge \neg A$$

further, employing classical logic it would be followed with a conclusion that:

$$T \cup \{D1, D2/D2'\} \vdash D3 \rightarrow A \wedge \neg A$$

which, according to the law of contraposition, results in:

$$T \cup \{D1, D2/D2'\} \vdash \neg(A \wedge \neg A) \rightarrow \neg D3$$

and as the theory includes an antecedent of the above implication:

$$T \cup \{D1, D2/D2'\} \vdash \neg(A \wedge \neg A)$$

then by the rule *modus ponens* it results in:

$$T \cup \{D1, D2/D2'\} \vdash \neg D3.$$

This conclusion means that:

($t1$) *It is not that a sentence is possible, but that neither now it is true nor it will be true in the future.*

---

180 According to Barnes, these rules relate to the logical reasoning of Philo, Diodorus and the Stoics. However, they can be also found in the works of Stagyrite. Barnes lists also other rules, which I shall leave aside, for they are provable on the basis of the two above-mentioned and the classical logic ([23], pp. 87–88).

181 This part of Diodorus' definition was supposed not to raise doubts in ancient times ([23], p. 88).

that is:

($t2$) *If a sentence is possible then it is now true or will be true in the future.*

what together with (R_t3) gives a substantiation for Diodorus' definition:
(*Def. of Possibility*): *A sentence is possible if and only if it is now true or will be true in the future.*

The above scheme of reasoning we attribute to the very reasoning of Diodorus Cronus. Every measure that shall specify set T and will express in its language the premisses $D1 - D3$ (with additional symbols for modalities) can be named the reconstruction of the reasoning of Diodorus Cronus, or simply **RDC**. Therefore, using this abbreviation does not have to mean a certain historical reasoning, carried out by Diodorus himself.

Even though this work has no historical ambitions, it is worth to mention that the original lost reasoning of Diodorus Cronus most probably also fell within the presented scheme, of course, as long as it had complied with the requirements concerning the specification of the set T, which, due to the indicated lack of relevant logical measures may seem highly problematic. Supposing that the Reasoning of Diodorus Cronus has been based on the inference from monosemous premisses (not e.g. with equivocation), even unknown, then their explication would have given us a solution to which logic we should extend the classical logic in order to reconstruct it accurately. We allow, however, various historical, perhaps incorrect and inaccurate, reconstructions.

The question of reconstruction carries two more important remarks. Notice, that if on the basis of theory T (and additional rules mentioned above) and the three premisses from the Reasoning of Diodorus we get a contradiction, then as long as the T along with the rules is not contradictory, the contradiction occurs on the basis of one of the premisses. *A priori* there are three possibilities. Either one of the premisses is self-contradictory or it results in contradiction in combination with the rest of the premisses, or in combination with the theory. The first possibility seems though excluded. For none of Diodorus' premisses are self-contradictory. There is also no suggestion that the premisses create a contradictory set of assumptions, i.e. on the basis of the classical logic a contradiction can be deducted from them. Contradiction though should occur by addition of a relevant theory, on the basis of which these three premisses lead to a contradiction. Besides, it is clear that the fact that Diodorus rejected the third premiss has been a result of his philosophical orientation, his polemists have been rejecting the other two. The presented scheme does not, however, exclude that a rejection of premiss $D3$ from the whole set protects it from a contradiction. Nevertheless, a contradiction may result from an addition of $D1$

or $D2/D2'$ to T, or the both together. Therefore, at each reconstruction premiss, $D3$ must play an important part in the reasoning, without which there would be no contradiction.

The next condition relates to the premisses. We shall use them in such form as was presented in the previous chapter, with the two variants of the second premiss and the two possible scopes of denotation of the word sentence. By all means, every time these premisses will be expressed in a language of the theory they will be added to. However, it will not always concern modal phrases 'necessary that', 'possible that'. In the logics, we shall employ, these phrases must occur, therefore, there will be no interpretation set in advance, even though they are present in each of the three premisses. Their meaning will be established only by **RDC** by means of purely syntactic transformations that will have enabled defining these phrases in a simpler terms of the given temporal logic, where **RDC** is being carried out. Therefore, these phrases should be used in conclusions as not interpreted phrases. It requires setting aside all connotations related to them.

5.3 The issue of *futura contingentia*

The issue of a logical value of the sentences about the future, resulting from applying the principle of bivalence to all the sentences without restrictions, is particularly well noticeable in reference to the sentences, in which case an intuition suggests that they are not characterised by any necessity resulting from the present states of affairs. Let us compare two sentences:

(1) *Tomorrow the sun will rise.*
(2) *Tomorrow I will go to the cinema.*

The first sentence seem to be based on certain physical and astronomical facts, observations and laws. Thus one could believe that such state of affairs, as that the sun will rise tomorrow, is really preordained, and the sentence should be considered true, as today all the sufficient physical conditions for the sun to rise tomorrow are met. In the second sentence, the occurrence of tomorrow's going out to the cinema is based, among others, on today's decision that may change many times until tomorrow. Thus the objective setting in today's states of affairs seems weaker than in the case of the first sentence. We do not claim, of course, that this classification is clear. However, it has been clear for the philosophers, who discussed the problem of *futura contingentia*, that is the future not yet determined from the human point of view. It can be though widened on all types of sentences, we have written about. For sentences such as (1) are not related to any logical necessity. Also, for any reason, the physical conditions could have changed until tomorrow and then the sun would not have risen — for this fact is not logically necessary.

The problem of *futura contingentia* troubled not only the ancient, but also posterior philosophers. It involves Christian philosophers in particular, as they have tried to correlate two, seemingly, contradictory concepts. On the one hand, the Christian doctrine allows free will, that is, also a possibility to make decisions and impact the shape of the future actions; and on the other hand, it ascribes the Creator properties of omnipotence and atemporality. God should thus know the decisions people will make and what will have been their consequences. Therefore, we should know a logical value of the sentences about the future, even though it depends only on the way the given objects will act[182].

The issue of God's omniscience is in a way similar to the problem of the *tomorrow's sea battle*. The attribute of God's omniscience can be considered as an ability that allows God to determine a logical value of each sentence from set $\mathbf{S_{TD}}$. If we limit the set of logical values to just two values, then God knows which sentences are true and which are false. This leads to analogical consequences like adopting the principle of bivalence for any sentences without referring to the divine knowledge. These approaches substantiate the stance that we have termed before as *quasi*-determinism. Thus, adding the additional modal premisses accepted by Aristotle, it can be claimed — as described in the chapter about the *tomorrow's sea battle* — that for each sentence either it is necessary or its negation is necessary.

In order to avoid the problem of determinism, systems of logic have been developed, which were not based on the principle of bivalence. Amongst many possible solutions, one of the most interesting and also the earliest is the multi-valued logic approach. Contingent sentences about the future can be assigned an intermediate logical value, just like proposed by Jan Łukasiewicz ([168, 169], [170], p. 278), motivated by the very problem of determinism and the principle

182 There have been various positions on this issue. It has been denied e.g. that both the dogmas were true, by rejecting human freedom; or claimed that a logical value of sentences about the future was not yet determined, thus they cannot be subject to God's knowledge. Some philosophers have attempted to correlate these two dogmas, either claiming that the divine perspective is atemporal and does not distinguish the past from the future, while people participate in time, or allowing different alternatives of the future that God knows about, and which people choose only then. The overview of these opinions can be found in the work of Øhstrøm P. and Hasle P. F. V. ([205], pp. 87–108).

of bivalence; that has been introduced to contemporaneous philosophical dispute by an article of Tadeusz Kotarbiński *On the Problem of the Existence of the Future* ([165], p. 47). Within this type of the logics of time, this approach has been developed, among others, by Artur Prior [230], Nicholas Rescher and Alasdair Urquhart [241], and, in Poland, by Kazimierz Trzęsicki [285].

Another way to avoid the deterministic consequences associated with the principle of bivalence is the development of temporal logic that, although does not reach beyond the two values, is not based on the law of excluded middle but arises through expanding the language and the axioms of intuitionistic logic. In literature, such logics are also considered indeterministic (cf. Trzęsicki K. [283], Surowik D. [266]).

We have, however, mentioned that the Ancients accepted both the principle of consistency and the law of excluded middle. Therefore they have approved, despite certain difficulties, the principle of bivalence. Hence, in further considerations only such logics should be employed that constitute a development of classical logic.

How to avoid, though, the problem of determined logical value of the sentences about the future, and at the same time not to determine the issue of the *quasi*-determinism? In the chapter on the concept of truth, we have drawn up some possible approaches to this issue. Among these, we have analysed an approach allowing different possible futures, when the same sentence carries different logical values. The idea that a determinist perceives the future in a branching way, comes from Saul Kripke and has been developed in various interpretations within the logics of branching time (Øhstrøm P., Hasle P. F. V. [205], pp. 180–196). At a certain chosen place, called the present, time can be

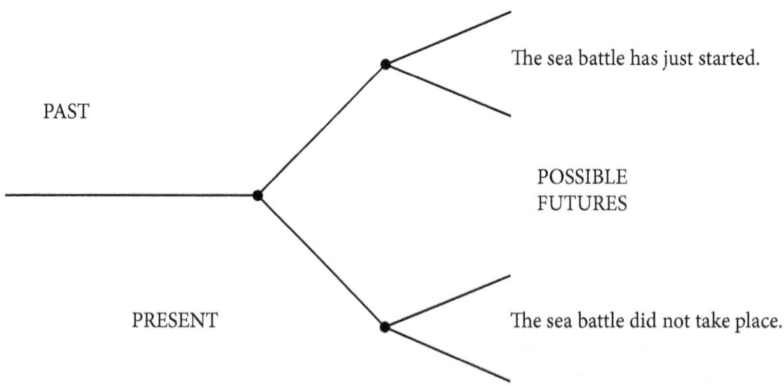

considered branching, i.e. allowing various futures, e.g. in one branch there is a sea battle, and in another one, for any reason, the sea battle does not occur. In each one, due to the course of events, there arise new possibilities, thus new branches. This way each sentence generates two new branches — one, where it is true, and the other one, where its negation is true. The following figure shall present such situation for the sentence 'There will be a sea battle tomorrow', which generates exactly one branch.

This example, along with the idea of branching time, reveals that the considerations about logical structures do not relate to the physical time. The individual branches of the future should not be regarded as similarly *real*, as e.g. the present. There are no sea battles fought there and no states of affairs occur there. They should be rather considered as sequences of possible worlds available in the future. Along with the passage of the history, when the point of present time moves forward, in the history of the world there will be exactly one branching amongst many available earlier, and sentences stating about the states of affairs from the past will become necessary, whilst their negations will become impossible. Hence the passage of thus comprehended time results in the loss of possibilities.

The described approach allows to apply the principle of bivalence to all the sentences from set S without an immediate consequence in the form of *quasi*-determinism, but then the sentences about the future carry a value related to the determined branches.

Naturally, in the study on the **RDC**, we will not presume time branching into the future. We will analyse, among other things, if the concepts, required to reach the conclusion that Diodorus was attempting to achieve, allow the structure branching into the future. Therefore, we are planning to take into account any fragment of the structure of time and analyse if the concepts employed in the conclusion, and the obtained Diodorus' definitions expressed by means of these concepts, do not exclude such future — due to a random but determined moment — branching.

If the theory of **RDC** allows branching, then it does not carry deterministic consequences. We obviously mean such branches, which are not substantially identical. That would mean that the theory of **RDC** allows different future histories of the world and does not carry deterministic features. Nonetheless, it is difficult to test in practice whether the branching possibilities differ materially, that is if the world does not double itself. The available logical means allow the analysis of the existence of such possibilities in a formal sense, and this is a necessary condition for the future to take place in various ways.

5.4 Research problem and the method used

Under this study, we have set a goal to analyse the Reasoning of Diodorus Cronus in the presented above extent and according to the comments contained therein. Therefore, we are going to present various reconstructions of **RDC** along with their formalisations. Due to the method employed, this study positions itself within the logical philosophy, that is philosophy cultivated by means of logical tools[183]. The main burden of this book lays in particular on the accurately characterised concept of time. It shall be defined in such a way that the problem of **RDC** could have been strictly analysed. Another important reason why we will have defined such concept of time is of the following nature.

The authors of a review work *Temporal logic. From Ancient Ideas to Artificial Intelligence* state in the introduction that the concept of time is actually non-definable by means of simple concepts, due to its *sui generis* nature. They believe that the knowledge about time can be enriched by means of interdisciplinary studies (of such distant fields as physics, linguistics or sociology). They can lead to a new synthesis of knowledge about time, however, not necessarily may give the definite answer for the question about the nature of time. The result of such interdisciplinary research on time is meant, among others, to stabilise a conceptual network and a language for a description of time. Its important part is meant to be, in turn, the logic of time dealing with the issue of transforming the information considering the time parameter (Øhstrøm P., Hasle P. F. V. [205], pp. 3–4).

In their opinion, the logic of time constitutes a formal framework serving both the analysis of information related to the time parameter and offering a formal ontology of time within the semantic layer. This framework can be employed in the analysis of different issues, in particular philosophical issues related to the concept of time. Hence, leaning towards the described opinion and certain philosophical tradition in the deliberation on **RDC**, we shall further introduce a formal concept of time. It shall enable an analysis of various reconstructions of **RDC** precisely within the perspective of temporal logics.

The aim will be to analyse different reconstructions of **RDC** — both the old ones and the newer ones — to determine the logical tools required to reach the justification for Diodorus' definition. Demonstrating that a formal reconstruction of **RDC** requires to employ a specified system of temporal logic, will allow to state unambiguously that it requires a specified logical time structure

183 Thus the aim for my work is to be spiritually close to the tradition of philosophical logic (cf. Perzanowski J., Pietruszczak A. [213]).

associated with the given logics. In particular, we shall be reflecting on whether given reconstruction allows, discussed in the previous chapter, time branching in the future, and thus whether the concepts of possibility and necessity, defined within it, are of a deterministic nature or not.

In the following chapters, we shall present the concept of time and an outline of the logical systems, we shall employ in the reconstructions, and also the axioms for different time structures corresponding to them.

6 Dates, tenses, sentences *vs.* time structures

6.1 Dates

When describing the empirical world by the means of the natural language, we provide the space-time location of the characterised states of affairs. For the states of affairs occur in time and space. The phrases we shall name — by expanding the colloquial meaning of word *date* — the dates, are being used to express the time parameter. By means of the adopted in practice calendar, dates allow to position the state of affairs described in a sentence in relation to other states of affairs occurring in the adjacent locations determined by means of other dates. Therefore, the calendar system and the order of the dates enable understanding one of the conditions that must be met for a sentence to be true, e.g. the sentence:

(1) *On 1st of September 1939, Germany invaded Poland, on 17th September Soviet Union invaded Poland.*

is true if, within the determined means of the time measurement, the days 1st and 17th of September 1939 refer to a certain time period when the Second World War broke out. Thus, the understanding of the dates and referring them to a certain metrological system constitutes the necessary condition of determining the logical value. For it is not possible to establish the time interval in which the given sentence locates the described state of affairs, without determining the metrological system the date belongs under.

Certain expressions, however, do not involve the dates of a metrological system but base on the time interval of their utterance, as e.g. a sentence:

(2) *The haymaking starts tomorrow.*

The phrase *tomorrow*, that also serves as the date, can be understood unambiguously only if the date denoting the utterance day is known. For it denotes the next day. Therefore, there are some phrases that resemble the dates, but actually bear the characteristics of occasional phrases.

The above remarks also indicate that the dates usually denote certain time intervals, not points of time. The sentences with such dates describe the states of affairs located in many points of time within the concept of time accepted in this study.

6.1.1 Dates and the *pseudo-dates*

Therefore, some of the dates refer to certain temporal objects, maintaining, within a determined metrological system, something that some authors refer to as the *chronological stability*, while the other expressions, e.g. *tomorrow*, are *chronologically unstable* (Rescher N., Urquhart A. [241], p. 27–30). The dates are the expressions that, within the determined metrological system, do not change their denotations regardless of the time they are uttered in, while the denotation of the *pseudo*-dates depends on the temporal context of the expressions. Thus, in the case of the sentences determined in time we are dealing with the dates, while the sentences temporally indeterminate contain at most *pseudo*-dates or there are no expressions denoting any temporal objects at all.

6.1.2 Denotations of the dates

The dates we use in the natural language refer to the whole time intervals. For the objects they denote can be divided into smaller temporal ones, e.g. a year consists of twelve months, months consist of a number of days, etc. Whereas, the points of the time structures established by us are to be considered the smallest temporal objects. Certainly these are not years, days, nor any long time intervals. For these can be divided into smaller temporal objects, while the points of time are indivisible. Of course it should not be preordained that the points of time are denoted by the smallest time units we can use to measure and express the time.

Knowing the benchmark of the established metrological system we can assume that e.g. a year, according to the set metrological system, denotes left-closed and right-open time interval $[t', t'')$ within the given time structure. Hence, the year starts at point t', whereas point t'' belongs to the next year[184]. Similar procedure can be followed regarding the smaller and bigger than year units, accordingly to the relationship between them. Such periods of time we shall call the time intervals. Thus, in a certain metrological system, the dates denote the time intervals[185].

---

184 If we worked with the unit corresponding to the point in time, we would be able to allow closed intervals $[t]$ that simply correspond to these units.

185 Such approach usually leads to considering the time structures defined by a set of time intervals, and the relations between a temporal precession and an inclusion defined on a Cartesian product of the set of the intervals (cf. Benthem J. F. A. K. [29]). Nonetheless, I am going to analyse **RDC** within the point structures without extending them into the intervallic structures. For the point structures belong

6.2 Grammatical tenses

Grammatical tenses function in many natural languages. Naturally, it is not our intention to express any general thesis on the relationship between the grammatical tenses and the use of language in its descriptive function[186]. We are also not going to analyse various grammatical tenses, as languages vary in this regard. Let us indicate just a few important aspects on the example of the present, past and future tenses in Polish[187]. Let us compare three sentences determined in time.

(3) *On 12th of December 2003 in Toruń, an art exhibition will take place.*
(4) *On 12th of December 2003 in Toruń, an art exhibition takes place.*
(5) *On 12th of December 2003 in Toruń, an art exhibition took place.*

These sentences differ only in terms of the grammatical tense of verb *to take place.* The linguistic practice suggests that this difference is associated with the location in time of a person that utters the individual sentences. In the first example, this person is located at the time prior to 12th of December 2003. The usage of the second sentence does not reflect the intuitions associated with the expression of the proposition on the state of affairs that is temporally coincident with the time of the utterance. Attempting to express such information, we would have rather said: *today in Toruń, an art exhibition takes place.* However, knowing that *today* denotes 12th of December 2003, without the change of information, we can assume that this sentence expresses the same proposition as (4). Sentence (5), analogously as (3), is structured by means of the adequate grammatical tense, in this case it is past simple, to express the fact that the state of affairs in question is past in relation to the time when a person expresses this information. The location in time of the person uttering each sentence is, therefore, undoubtedly different.

---

to some of the oldest and most fundamental measures of the formal representations of the time (cf. Allen J. F. [9]).

186 For there are many works on this subject that approach this problem from different angles (e.g. Bennett M. [30], Kuhn S. [141], Saarinen E. [252]).

187 However, these remarks do not apply exclusively to English language; on the basis of its parts of the grammatical tenses, the presented hereafter so-called *tense logic* based on similar assumptions, has developed. These assumptions — as the authors claim — do not apply to all the grammatical tenses (cf. Bäuerle R. [26]) nor to all types of the statements that feature the entangled grammatical tenses, e.g. in the structures of reported speech (cf. Higginbotham J. [106, 107]). However, I shall consider simple tenses in relation to the beforehand defined simple sentences, presenting the presumptive iterations of the tenses at the logical operators level.

With a certain lack of style, these sentences can be transferred into the sentences expressing the same information in a clearer manner regarding the person uttering them. One of the suggestions comes from the work of Robert McArthur [180] and results in disposing of the past and future grammatical tenses in the following neutral paraphrases:

(6) *On 12th of December 2003 in Toruń, an art exhibition takes place and the 12th of December 2003 is later than now.*
(7) *On 12th of December 2003 in Toruń, an art exhibition takes place and the 12th of December 2003 is now.*
(8) *On 12th of December 2003 in Toruń, an art exhibition takes place and the 12th of December 2003 is earlier than now.*

Such manner of formulating the considered sentences allows to understand that grammatical tenses actually serve to i.e. determining the temporal position from which the sentence is being uttered, in relation to the state of affairs expressed by it. Agreeing that the presented paraphrases express the same information as the initial sentences, it turns out that the utterance of e.g. sentence (3) later than on 12th of December 2003 expresses a false proposition due to the second segment of the paraphrase (6). The second segment of the conjunction of the above paraphrases plays an important part in expressing the time in which the truthness of sentence 'On 12th of December 2003 in Toruń, an art exhibition takes place.' is being stated[188]. Therefore, if sentence (6) is true, then it means that sentence 'On 12th of December 2003 in Toruń, an art exhibition takes place.' is true before 12th of December 2003.

In this simplified model, in the case of the sentences determined in time, the use of e.g. future grammatical tense could have been understood as the acknowledgement that the given sentence stating something about the future time interval is true in the present time interval. Thus, the issue of the logical value of the sentences determined in time stating about the future could have been analysed without the reference to the grammatical tense, and only by considering them in respective time interval, earlier than the stated by them states of affairs are supposed to take place. In this situation and with such understanding of the functions of the grammatical tenses, they can be left aside.

---

188 This consequence may appear strange, but, on the other hand, modern users of Polish language would have considered he sentence 'The Battle of Grunwald will take place in 1410, near Grunwald.' as, at best, false or, in a worse case — absurd. The meaningful and at the same time true sentence would have read as follows: 'The Battle of Grunwald took place in 1410'.

And yet, the grammatical tenses play an important role in analysing such reconstructions of **RDC** where the sentences temporally indeterminate are used. For they instantiate not only the temporal location of the expression, but also, to a certain extent, indicate the time interval in which given sentence is to be true. Employing the above examples, but without the dates, we can, not too elegantly but in a quite expressive manner, reveal the grammatical tenses of the individual sentences as follows:

(9) *It will be so that in Toruń an art exhibition takes place.*
(10) *It is so that in Toruń an art exhibition takes place.*
(11) *It is so that in Toruń an art exhibition took place.*

In this case, it is not possible to carry out such a paraphrasing procedure as performed in the previous case, without a change of the meaning or a stylistic bizarreness. However, knowing the point of time in which their usage takes place, we also know a certain time range they refer to. In sentences (9) and (11), an art exhibition in Toruń is to take place in a later or earlier time interval than a point of time or time interval they are uttered, thus later or earlier than the *present*[189]. Thus, in sentences (9) and (11), the grammatical tenses should be considered operators that, due to a certain specified present, instantiate the time in which in Toruń an art exhibition takes place. Besides, the situation is similar regarding sentence (10), that refers to the *present*. Assuming the given point present, these sentences can be considered stating the propositions that in some future, or respectively past, in relation to the present an art exhibition takes place in Toruń[190].

Thus, in the case of the sentences undetermined in time, it is required to set the points of time for the statements, not only to allow considering the issue of their logical values, but most of all to comprehend their meanings, for at the different points these statements express different meanings. As for the

189 The present time, comprehended within the point time structure, undermines the colloquial linguistic intuitions and the phenomenological and psychological concept of the present. Within the point concept of time, *the present* is basically a single distinguished point, while in a natural language by saying *I am working now*, I can express that I have been working for an hour now (cf. Kałuszyńska E. [128], pp. 144–145).

190 This approach is consistent with the outline of semantics proposed for the natural language by Ernest Lepore and Kirk Ludwig in their work [154] and in general with the semantics of so-called tense operators.

sentences determined in time, their meaning is known — what is required to determine their logical value are the non-linguistic conditions.

In the further works, we shall employ both of the approaches to the question of the grammatical tenses, omitting them at all in the case of the sentences determined in time, or, in the case of other sentences, considering the tenses the operators that refer the sentences to the due directions of history.

6.3 Logical values of the sentences within the time structures

Since the dates denote the time intervals, this characteristic naturally arises the question about the relation of the logical value of the sentences to the time structures. For within the time structures described below, there is no formal concept of time interval. Let us consider the following example. Assuming that in a point of time that belongs to a certain time interval denoted by date 1st of July 1999, there occurs state of affairs *s*. Then, according to the definition of truthness (2.3.1, p. 48), the following sentence is true:

(12) *In July 1999 there occurs state of affairs s.*

Let us though examine this example from the other angle. Let us assume that (12) is true. By that definition, state of affairs *s* should belong to the history of the world in July 1999. Does it mean, though, that it should occur in every point of time within the time interval denoted by date July 1999, or only in certain ones? Let us consider another, more accurate example.

(13) *I worked today.*

The truthness of this sentence may be based on the fact that, on the given day it is being uttered, a person uttering it works for the whole day; i.e. in each point of time that belongs to the time interval denoted by phrase *today*, e.g. $I = [t', t'')$, it is so that I work, i.e.

(14) *In each point of t from time interval I it is so that I work.*

It seems rather difficult to understand, because the sentences e.g. the following one:

(15) *Today I worked for 0,0000000001 of a second.*

give an impression of a rather humorous approach to some other situation dependent on the context, rather than the expression of the proposition that in such a period of time I have worked. Therefore, the truthness of the majority of sentences such as (13) cannot be reduced to the generalised conjunction of the sentences stating about the occurrence of this state of affairs in each point of the time interval[191]. Of course, there are examples that do not allow to generalise this remark:

(16) *In July 1999, Toruń is a city.*

In this case, the fact that Toruń is a city in July 1999 means that in each point of time belonging to the time interval denoted by the date *July 1999* Toruń is a town. For such sentences as (16), we can always provide a certain set of more detailed sentences that when put together are equivalent to it.

The described problem results in the confusion whether the truthness[192] of a sentence can be represented by the set of some less general — in terms of the time parameter — sentences. For the following principle:

> *Any sentence* **s** *from set* $\mathbf{S}_{\mathrm{TU}}$, *supplemented with the name of time interval* $[t', t'')$, *is true if and only if sentence s is true in each point of time of this interval.*

is contradicted by examples (13), (14). The truthness of such sentences as (13) could have probably been represented by various sentences describing states of affairs from individual points belonging to interval *today*. Here resides some silent assumption that the states of affairs, e.g. the work for the whole day, can be divided into simpler, in temporal sense, states of affairs that even though they are not the work itself over the whole time interval result in such states of affairs as the work for the whole day[193]. Let us illustrate it with the following example describing the simpler states of affairs:

(17) *At 17:00:01 on 1st of July, Jan entered the shop.*
(18) *At 17:00:05 on 1st of July, Jan asked for a kilogram of sugar.*
(29) *At 17:00:10 on 1st of July, Jan asked for a kilogram of flour.*

---

191 This problem in terms of English language has been indicated in a work of Vendler Z. [286]. The subject of the distinction between so-called homogeneous predicates, that is such which hold over the given time interval only if they hold over each point of the interval, and non-homogeneous predicates, that is such that e.g. hold over the given time interval, but do not hold over each point of it; is, however, widely considered within the formal representations of time for so-called intelligent systems (cf. Allen J. F. [9]).

192 I omit herein the question of being false, as the problems it causes are analogous.

193 This issue has been considered in a quite interesting way by e.g. J. F. A. K. Benthem. These considerations are apparently close to the Zeno's paradoxes. The analysis of the states of affairs or events, leading to a certain outline of the theory of such objects, thus to the ontology ([29], pp. 113–124). In my work I shall employ the suggested simpler and — in the long run — unsatisfactory solution.

(20) *At 17:00:15 on 1st of July, Jan asked for a kilogram of sugar.*
(21) *At 17:00:45 on 1st of July, Jan paid for the sugar and flour.*
(22) *At 17:01:05 on 1st of July, Jan left the shop.*

The truthness of these sentences results in the truthness of the following sentence:

(23) *On 1st of July Jan bought some sugar and flour in the shop.*

but it does not have to be the other way round. The truthness of sentence (23) may be based on various combinations of the states of affairs that occurred on 1st of July and resulted in such a state of affairs that Jan bought some sugar and flour. However, if the logical value of such partial sentences is determined, then the logical value of such sentences as (23) is determined as well. Therefore, in the further part we shall claim that while indicating that certain facts occur for sentences referring to the points, we simultaneously imply the same characteristic of the sentences stating about the greater time intervals that are equivalent to or at least result from certain sets of sentences referring to the points of this interval.

Therefore, we shall be employing the sentences from set **S** assuming that by defining the sentences about the intervals by means of the sets of the sentences about the points, does not detract the generality of the considerations. Thus, stating that a sentential variable ranges over a set, e.g. $\mathbf{S}_{\mathbf{TD}}$ we will mean the sentences describing the states of affairs in the given point of time, whereas for the sentences describing the states of affairs in the multi-point intervals, we shall interpret as expressible by the means of the former.

The further works shall be carried out by means of the described simpler interpretation of the grammatical tenses and the simple model of perceiving and determining the logical values from set **S**.

7 Logic

In the present chapter, we shall proceed to more technical issues. We will introduce the appropriate formal tools that enable the precise analysis and representation of the individual reconstructions of **RDC**. We will briefly discuss the two types of the temporal logics: the tense logics and the chronological logics. Naturally, we are not going to describe them comprehensively, but only to the extent they will have been employed in the last part of the book. Therefore, let us present the principles of their language construction, carrying out proofs on their basis, and the semantics. In the second part of this chapter, we will present the fundamental for the analysis of **RDC** concept of logical structures of time, that is, the so-called *frames* of the passage of time.

We shall be employing logical tools that constitute the extension of the Classical Propositional Logic (for short **CPL**). It means that the presented logics will respect all the rules of inference in accordance with **CPL**, and some of them may also include the language of **CPL**.

Herein, we shall not remind the mechanism of **CPL**, assuming that the reader possesses a relevant understanding of this subject. We will only mention certain most important facts and a notation adopted in the study.

The set of the following functors belongs to the alphabet of the language of the classical logic: $\{\neg, \wedge, \vee, \rightarrow, \leftrightarrow\}$, respectively: negation, conjunction, disjunction, implication and biconditional; in the classical, boolean interpretation. Additionally, to the alphabet of the language of **CPL** also belongs the following countable set of the sentential letters: **Var** $= \{p_1, q_1, r_1, p_2, q_2, r_2, \ldots\}$ and brackets), (.

The expressions of the language of **CPL** are finite sequences of the symbols called formulas and formed in a following standard manner: $\neg A$, $(A \wedge B)$, $(A \vee B)$, $(A \rightarrow B)$, $(A \leftrightarrow B)$, and letters from **Var**. We assume that letters A, B, C, … are metavariables ranging over the set of the formulas.

The interpretation of the language of **CPL** is any function V that assigns a logical value from set $\{0,1\}$ to each formula so that it retains the boolean interpretation of the functors which constitute the complex formulas:

1. $V(\neg A) = 1$ if and only if (for short: iff) $V(A) = 0$
2. $V(A \wedge B) = 1$ iff $V(A) = 1$ and $V(B) = 1$
3. $V(A \vee B) = 1$ iff $V(A) = 1$ or $V(B) = 1$
4. $V(A \rightarrow B) = 1$ iff $V(A) = 0$ or $V(B) = 1$
5. $V(A \leftrightarrow B) = 1$ iff $V(A) = V(B)$.

The given inference, that is, to draw conclusion S_{n+1} from the set of premisses $\{S_1, \ldots, S_n\}$, can be stated true iff for any interpretation V, if $V(S_1) = 1, \ldots, V(S_n) = 1$, then $V(S_{n+1}) = 1$. **CPL** allows to differentiate between the valid and invalid inference schemes. The valid inference schemes are based on the meaning and distribution of the Boolean functors and the sentential letters occurring in the sentences, that is, on the structures of the formulas.

In the reconstructions of the reasoning, whilst employing the temporal logics, we shall sometimes refer to the proof transformations based solely on the meaning of the boolean functors, that is, to the inferences valid in the light of the classical logic. In proofs, this fact will be indicated by placing the numbers of the premisses on the right side of the proof and adding symbol: **CPL**.

In some of the further-described logics, there arises the necessity to quantify over the points of time or of distance between them, or, possibly, over some other objects. Therefore, there are quantification rules required; these are provided by the First Order Logic. Let us remind a few concepts related to **FOL** and determine certain convention.

The family of disjoint sets constitutes the alphabet of **FOL**. It includes: a set of individual variables $\{x_1, y_1, z_1, x_2, y_2, z_2, \ldots\}$, a set of individual constants $\{a_1, b_1, c_1, a_2, b_2, c_2, \ldots\}$, a set of predicate constants $\{p_1, q_1, r_1, p_2, q_2, r_2, \ldots\}$, a set of functional constants $\{f_1, g_1, h_1, f_2, g_2, h_2, \ldots\}$, a set of logical functors $\{\neg, \wedge, \vee, \rightarrow, \leftrightarrow, \exists, \forall\}$, classical and the existential and universal quantifiers, and also brackets), (. There are certain functions associated to the sets of the predicate and functional constants; they assign to the elements of the sets the natural numbers greater than zero, indicating the quantity of the arguments the given predicate or the functional constant involves.

The finite sequences of the symbols, called terms, constitute the auxiliary expressions in the language of **CPL**. The individual constants and variables are the terms; moreover, the terms are also structured as follows: $\alpha(t_1, \ldots, t_n)$, where α is the n-argument functional constant and $t_1, \ldots, t_n$ are the terms.

The expressions of the language of **CPL** are the finite sequences of the symbols called formulas and formed in a standard manner: $\beta(t_1, \ldots, t_n)$, where β is a n-argument predicate constant, whereas $t_1, \ldots, t_n$ are terms, and $\neg A$, $(A \wedge B)$, $(A \vee B)$, $(A \rightarrow B)$, $(A \leftrightarrow B)$, $\exists_y A$, $\forall_y A$, where A and B are formulas, and y is an individual variable.

As you can see, the language of **FOL** is more complicated than the language of **CPL**. This allows to express the subtler relations occurring within the inferences than merely the relations based on the understanding of the boolean constants. For **FOL** describes also the relations occurring between the quantifiers and the other constants. Due to this complication, we shall not introduce herein the concept of the interpretation for the language of **FOL**.

The concept of correct inference is still based on the idea that, for any interpretation, with the true premisses also the conclusion is true. In the cases of some reconstructions — where in the language, apart from boolean constants, there are also quantifiers — the rules of the correct inferences in accordance with **FOL** will also be employed. In the parts of the proof, where transition from one proof step to another is based on the inference scheme valid for **FOL** — as in the case of **CPL** — on the right side of the proof we shall place the numbers of the premisses and **FOL**.

In the case of the application of **CPL**, as well as **FOL**, in the proof transitions, we shall sometimes write down the indirect proof steps instead of the direct steps. This may by helpful in the analysis of the proof correctness.

7.1 Temporal logics

By temporal logics, or logics of time, we mean the logics that aim to systematise the reasonings considering the relations between the sentences built up from functors subordinating their logical value from the time parameter. It can be illustrated with the following example:

(1) *Jan goes to school.*
(2) *It will always be the case that it has some time been the case that Jan goes to school.*

Even though there is no logical relation based on the classical functors between these two sentences, but, from the temporal logic's point of view, sentence (2) results from sentence (1). The concept of the logical consequence is, however, referred to the interpretation of specific logical constants that guarantee the above relationship. This case is about the logical constants representing different temporal expressions. Provided the classical logic is a theory of the classical functors, the temporal logics are theories of the temporal functors.

The inception and development of the temporal logics varied in motivations. In ancient times, the temporal interpretations of the necessities and possibilities have been considered offering different solutions depending on the school. Diodorus Cronus opted for the temporal definition of the implication and also placed his famous argument concerning the relationship between the sentences and the time. Also in the Middle Ages, the problem of time in reasoning[194] has

194 The extensive history of the relationship between the problem of time and the development of the logic, especially the early one, is quite comprehensively discussed in the work *Temporal Logic. From Ancient Ideas to Artificial Intelligence* [205]. However, the work incorrectly omits the contribution and the priority of Jerzy Łoś.

been taken into consideration. But only nowadays, along with the emerging of the formal logic in the form of the logical calculus, the problems formulated and considered by preceding philosophers and logicians could have been formulated in a more clear manner. That is the very reason why amongst the incentives for the creation of the modern temporal logics one can name the historical cause related to the attempt of interpretation, comprehending and systematising the ideas of the preceding theorists (cf. Prior A. [229], ch. 7). The basics of the logic of time emerged as well for the use of the methodological analysis of the inductive approach in science (cf. Łoś J. [166]), as well as a result of the analysis of grammatical tenses (cf. Reichenbach H. [238]). The suggestions of Jerzy Łoś have been further developed by Arthur Prior [230] and by Nicholas Rescher [240]. The logic acknowledging the time parameter has been entirely independently constructed by von Wright G. H. [298, 299]; this logic is called (*the logic of change*). Nowadays, the temporal logics in the forms of tense logics are developed based on the temporal functors. Apart from the historical and philosophical incentives for their development, there are also mentioned the linguistic considerations related to *computer science*, and even mathematical ones (Burgess J. P. [44], pp. 94–96).

In the study on the relationship between **RDC** and logical structures of time, while analysing the proposed interpretations, we shall employ two types of logics. The first ones are formed by enriching the language of the classical logic with one operator binding the logical value of the sentences with the respective points in time. The other type of the logics emerges based on several other additional functors. We shall begin with discussing the first one.

7.1.1 Temporal interpretation of the positional logic

The simplest models of the logic, that consider the statements employing temporal expressions, are the temporal interpretations of the positional logic[195].

Within these logics, the language of the classical logic (or another one, upon which the positional logic is superstructured, e.g. many-valued logic) is enriched with paramaterising operator $\mathcal{R}_i$ that, from a syntactic point of view, creates a sentence from a sentence, thus being a sentence-forming functor of a sentential argument. Index i indicates the position the sentence, from within the range of the operator, is referred to[196]. The sentences built by means of operator

---

195 Also called, by Rescher and Garson, the topological logic [81].

196 The realisation operator can be characterised as $\mathcal{R}_i$ — stating that $\mathcal{R}$ refers the sentence to position i — equally as well as the very $\mathcal{R}$ associating the position

$\mathcal{R}_i(A)$ can be decoded as: *sentence A is realised in position i*. The measure of decoding should be chosen accordingly to the nature of the parameter. For the set of the locations, the sentences referred to should be interpreted differently — as a set of spatial locations, possible worlds, the numbers constituting solutions to equations, the geometry in which the given theorem is true or false, etc., or, finally, as a set of the points of time ([81] p. 537, p. 545). For example, the following sentence:

(3) $\mathcal{R}_3(x = 7 - 4)$

could be decoded as 3 *is a solution of the equation* $x = 7 - 4$[197], whereas sentence:

(4) $\mathcal{R}_i$(Socrates is sleeping)

for value i corresponding to e.g. a room in Socrates's house, as 'sentence *Socrates is sleeping* comes true in the room in the Socrates's house', or more generally, 'sentence *Socrates sleeps* comes true in location i'.

As you can see, the values of the parameter (for operator $\mathcal{R}_i$ parameterisates sentences) may be various objects. The index of the operator can be a constant preserving the reference to a single object, or a variable adopting various references. Since the index can be considered a variable, then, apart from the laws of the propositional logic, subsequent to an adequate enhancement of the language, the laws of the logic of quantifiers are also figured in the positional logic. As a result, we get the following formulas $\forall_i \mathcal{R}_i(A)$ and $\exists_i \mathcal{R}_i(A)$ that respectively are read *for each i formula A is realised in i* and *there is such i that formula A is realised in i*.

The logics structured in such language can also have the temporal interpretation of the location. Following Rescher, we shall call such logics the chronological logics.

We will present a certain set of axioms and principles of these logics. In the indices, rather than any variables, we shall use letter t, alternatively with an index, that refers to the elements from the set of the locations in time. The presented axioms constitute some inspiration for the outline of the metric chronological logics we shall present in one of the following subchapters.

---

with the sentence. In practice, we shall employ both notations having in mind the same operator of realisation. The idea of the operator was invented by Jerzy Łoś [166]. The symbol for realization operator was introduced by Nicholas Rescher.

197 I am not willing to read this notation as *equation* $x = 7 - 4$ *is true for number* 3, even though such a way of reading seems natural. I am not willing to, for the word *true* I would leave for the semantic of the adumbrated logics. The sentence within the range of operator $\mathcal{R}_i$ is true only in a certain interpretation of the expressions occurring in it.

Therefore, let us proceed to the axiomatics of the chronological logics (cf. Rescher N. [240], Garson J., Rescher N. [81]). All these logics are based on the following specific rule:

$$\frac{\vdash A}{\vdash \mathcal{R}_t(A)}.$$

This rule states that for any formula A, if it is a theorem of given logic, then formula $\mathcal{R}_t(A)$[198] is also the theorem.

Additionally, in the axiomatics of Jerzy Łoś [166], there are the following axioms (apart from the special axioms that do not describe the actions of operator $\mathcal{R}_i$, but only the domain of the parameter)[199]:

1. $\mathcal{R}_t(\neg A) \leftrightarrow \neg\mathcal{R}_t(A)$
2. $\mathcal{R}_t(A \to B) \to (\mathcal{R}_t(A) \to \mathcal{R}_t(B))$
3. $\forall_t \mathcal{R}_t(A) \to A.$

The first axiom states that the truthness of the negation of the sentence at the point of time is equivalent to the falseness of this sentence at this point. This axiom expresses the principle of bivalence. For it consists of two material implications:

$$\mathcal{R}_t(\neg A) \to \neg\mathcal{R}_t(A)$$
$$\neg\mathcal{R}_t(A) \to \mathcal{R}_t(\neg A)$$

that are equivalent by virtue of the very **CPL** to the following schemes:

$$\neg(\mathcal{R}_t(\neg A) \wedge \mathcal{R}_t(A))$$
$$\mathcal{R}_t(A) \vee \mathcal{R}_t(\neg A)$$

that overall results in that at each moment t sentence either is or is not realised. For its negation is being realised.

The next axiom intuitively reflects the relationship between the material implication and the new operator: if at any point of time the given implication and its antecedent are implemented, then its consequent also is. The last axiom states that if the given sentence is realised at each point of time, then it can be accepted without referring to any point.

---

198 Sometimes, this rule is understood in such a way that its premiss states that formula A is the tautology of **CPL**.

199 It is worth to emphasise — for it is often omitted in the literature — that it was Jerzy Łoś who created the first chronological logics, and by that has been one of the creators of temporal logics in general.

The systems of axioms of Arthur Prior [230] partly constitute an extension of the axioms of Los. In the first system, the following axiom is added to the presented ones:

4. $\mathcal{R}_{t_1}(\mathcal{R}_{t_2}(A)) \rightarrow \mathcal{R}_{t_2}(A)$,

it states that the iteration of operator $\mathcal{R}_i$ is empty, for only the first operator influences the sentence.

In the second system Prior adopts the first two axioms of Los, whilst the third one he replaces with the following axiom:

3. $\mathcal{R}_n(A) \rightarrow A$.

Its meaning is based on the contextuality of phrase *now*. Letter n in the parameter of the operator means exactly *now*. Since it is an occasional phrase, the sentence realised in *now* is realised regardless of its position in time. Within this group of the axioms, the iterations are dealt with in a different manner than within the previous group:

4. $\mathcal{R}_{t_1}(\mathcal{R}_{t_2}(A)) \rightarrow \mathcal{R}_{t_1+t_2}(A)$.

For in this case, it is assumed that the positions in time are represented by numbers that indicate the distances between the positions from a certain starting point. If in t_1 a sentence is realised, and it states that in t_2 sentence A is realised, then sentence A is realised in $t_1 + t_2$ units from the starting point. The last axiom states about the relationship between the operator, which parameter is related to a quantifier, and another operator:

5. $\mathcal{R}_{t_1}(\exists_{t_2}(\mathcal{R}_{t_2}(A)) \rightarrow \exists_{t_2}(\mathcal{R}_{t_1}(\mathcal{R}_{t_2}(A))$.

Let us analyse the axiomatics of Rescher [81]. In the first one, to the 1st and 3rd Los's axioms, Rescher adds, fortified to equivalence, Prior's 4th axiom from the first group, and the 5th one, from the second group. The new axiom herein is the distributivity of operator $\mathcal{R}_i$ in relation to the following conjunction:

2. $\mathcal{R}_{t_1}(A \wedge B) \leftrightarrow (\mathcal{R}_{t_1}(A) \wedge \mathcal{R}_{t_1}(B))$.

In the second group, the 4th axiom is replaced by its metric equivalent. The presented axiomatics form a certain family of the chronological logics that, depending on the conditions imposed on the iteration and the position of *now*, require different interpretations.

The minimal system of logics with operator $\mathcal{R}_i$, in which the language does not allow nesting and the operator is distributed over the boolean constants (i.e. $\mathcal{R}_i(A * B) \leftrightarrow (\mathcal{R}_i(A) * \mathcal{R}_i(A))$ where $*$ is a binary boolean functor, and also the 1st scheme axiom of Łoś is the scheme of the theses of the system), has been defined in the work [119]. And even weaker systems, modelled for the description of time called the empirical time, has been defined by Marcin Tkaczyk [280]. The most comprehensive elaboration of the fundamentals of positional logic is book *Normalne logiki pozycyjne* (*Normal positional logic*) [121].

7.2 Time for the point moments

The time structures as well as the other concepts required for the analysis of **RDC** shall be defined by means of the formulas of the language of the first order **CPL** with the equality, of the standard notation (using symbol = as a denotation of the identity predicate), and of the basic concepts of the set theory. In particular, the binary relations defined on some set X will be considered (marked with capital letters: P, Q, R, S, …) as the subsets of Cartesian product X^2, the fact that a certain ordered pair $\langle x, y\rangle$ belongs to given relation R shall be denoted in one of the two following manners: $\langle x, y\rangle \in R$ or xRy. Additionally, the conversion of relation R will mean the relation denoted as R^{-1} that contains the pairs converse to the ones belonging to R. The equivalence relations, that is, the functions, will be denoted by lower case letters: f, g, h, … And characters $\in$, $\subset$, $\subseteq$, $\varnothing$ will respectively mean *to be an element of a set, proper inclusion, inclusion* and *empty set*. Moreover, since the relations are considered sets, there can be regular algebraic operations carried out on them, such as union, cut, complement of sets. We shall also put in that expression $x \neq y$ will have been considered a shortened version of formula $\neg(x = y)$. They both negate $x = y$.

Now we shall proceed to the presentation of the necessary concepts and the basic facts.

Definition 7.2.1 *The identity relation $I \subseteq X^2$ is a relation that meets the following condition:*

$$\forall_{x,\, y \in X}\,(xIy \leftrightarrow x = y).$$

Definition 7.2.2 *$R \subseteq X^2$ is a partial order relation iff for any x, $y \in X$ it meets the following conditions:*

1. *xRx — reflexivity*
2. *$xRy \wedge yRx \rightarrow x = y$ — antisymmetry*
3. *$xRy \wedge yRz \rightarrow xRz$ — transitivity.*

Subsequently, the partial order relations will be denoted with the following symbol: $\leq$, pair $\langle X, \leq\rangle$ will be called *partial order*, and set X a *partially ordered* set. If relation R meets any of its conditions, e.g. the reflexivity condition, then we will state it is *reflexive*; if it meets e.g. antisymmetry condition, then we will state it is *antisymmetric* etc.

Notice that by the concept of the partial order, the following fact occurs[200].

Fact 7.2.3 *Let $\langle X, \leq \rangle$ be a partial order. Then:*

$$\forall_{x,y \in X} (x \neq y \rightarrow (x \leq y \rightarrow \neg y \leq x)).$$

In turn, for the reflexive relations the following fact is true:

Fact 7.2.4 *If $R \subseteq X^2$ is a reflexive relation, then $\forall_{x,y \in X} (xIy \rightarrow xRy)$.*

Sometimes, whilst defining the time structures, a relation that includes only different moments can be useful. By that we mean the relation of strict order.

Definition 7.2.5 *$R \subseteq X^2$ is a relation of strict order iff for any x, $y \in X$ it meets the following conditions:*

1. *$\neg xRx$ — irreflexivity*
2. *$xRy \wedge yRz \rightarrow xRz$ — transitivity.*

Fact 7.2.6 *If relation $R \subseteq X^2$ is transitive, than the fact that it is irreflexive is equivalent to the following condition, called an asymmetric condition:*

$(*)$ $\forall_{x,y \in X} (xRy \rightarrow \neg yRx)$.

It means that the strict partial order relations could have been alternatively defined by condition $(*)$. Subsequently, the strict partial order relations will be denoted by symbol: $<$, pair $\langle X, < \rangle$ called *a strict partial order*, and set X *a strictly partially ordered set*.

Amongst the hereinafter generally defined time structures, we shall focus solely on the partial orders, only sometimes making use of the strict partial order relations and perhaps imposing additional conditions on the ordering relation. One of the fundamental conditions is a totality condition that along with additional conditions guarantees a straight line geometry of the modelled time.

Definition 7.2.7 *$R \subseteq X^2$ is a relation of linear order iff it is the partial order relation, and also for any $x, y \in X$ it meets the following condition:*

$(**)$ $xRy \vee yRx$.

---

200 The obvious facts will be left without proofs; more complicated proofs of the facts, lemmas and theorems will be presented or just outlined.

Condition $(**)$ by virtue of the fact 7.2.4 in the case of partial order relation is certainly equivalent to the following formula that expresses the totality condition:

$(**)'$ $\forall_{x,\,y\in X}\,(xRy \vee yRx \vee x = y)$.

In the further part of the study, with the use of symbol $\leq$, the condition from the definition shall be understood as follows: $x \leq y$ will mean that x is *identical to* y, or x is *prior to* y.

If $\leq$ is a linear order relation, then pair $\langle X, \leq\rangle$ will be called *a linear order*, and set X *a linearly ordered set* or *total*.

Definition 7.2.8 *$R \subseteq X^2$ is a strict linear order relation iff it is a strict partial order relation and also for any x, $y \in X$ it meets the following condition:*

i) $xRy \vee yRx \vee x = y$ *— totality.*

If $<$ is a strict linear order relation, then pair $\langle X, <\rangle$ we will call *a strict linear order* and set X *a strictly linearly ordered set* or *total*.

Sometimes instead of saying that given set $\langle X, R\rangle$ is e.g. a linearly ordered set, we will say that structure of $\langle X, R\rangle$ is *linear* or *strictly linear* etc.

An interesting and important for the further considerations property of the strict and partial order relation is their mutual definability based on the identity relation.

Fact 7.2.9 *(i) Let $\leq \subseteq X^2$ be a partial order relation, then relation R defined by the following condition is a strict partial order relation:*

$$\forall_{x,\,y\in X}\,(xRy \leftrightarrow x \leq y \wedge \neg x = y).$$

(ii) Let $< \subseteq X^2$ be a strict partial order relation, then relation R defined by the following condition is a partial order relation:

$$\forall_{x,\,y\in X}\,(xRy \leftrightarrow x < y \vee x = y).$$

The above fact guarantees that each partial order relation can be defined by means of the identity relation and the strict order relation, and also that each strict order relation can be defined by means of the identity relation and the partial order relation. Obviously, these facts can be supported. Analogously, the strict linear order relation can be defined by means of the linear order and *vice versa*.

Due to the above facts, whilst considering partially ordered sets, we will be using symbol $<$ meaning the elements that belong to relation $\leq$, but are not identical.

Definition 7.2.10 *We shall say that* $R \subseteq X^2$ *is a symmetric relation iff it meets the following condition:*

$$\forall_{x,y \in X}(xRy \rightarrow yRx) \text{ — } symmetry.$$

According to fact 7.2.3, it means that for the different objects belonging to relation $\leq$ it is not a symmetrical relation.

On the basis of the above concepts, we will now present what will be meant by time.

Definition 7.2.11 *The time or in other words the time structure is a relational system in the following form* $\langle T, \leq \rangle$, *where* T *is an nonempty set, and* $\leq$ *is a binary precedence relation defined on this set, that meets the following conditions:*

$i)$ $\forall_{t_1,t_2,t_3 \in T}(t_1 \leq t_3 \wedge t_2 \leq t_3 \rightarrow t_1 \leq t_2 \vee t_2 \leq t_1)$
$ii)$ $\forall_{t_1,t_2 \in T} \exists_{t_3 \in T}(t_3 \leq t_1 \wedge t_3 \leq t_2)$.

The elements of set T *will be called moments of time, points of time or, simply moments or points.*

Set T, as well as relation $\leq$ are thus nonempty. Ergo, the time domain includes a certain number of the moments. Apart from that, we do not assume anything neither about its cardinality nor about the nature of its elements. T can be a finite set as well as an infinite one, countable as well as uncountable, wherein of cardinality rather not greater than the cardinality of the set of real numbers — although the definition alone does not exclude more numerous domains.

Relation $\leq$ is to express the intuitions related to the directional passage of time. In practice, we will introduce a certain convention of reading it. Despite that $\leq$ allows the fact that $\langle x, x \rangle \in \leq$, we will anyway call it a relation of *being earlier than*. Therefore, about any objects x, $y \in T$, such as $\langle x, y \rangle \in \leq$, we will say that x *is earlier than* y. Hence its conversion $\leq^{-1}$ should be understood as a relation of *being later than* and accordingly, any objects x, $y \in T$, such as $\langle x, y \rangle \in \leq^{-1}$, should be considered x *being later than* y. Conversion $\leq$ can also be denoted as $\geq$.

Fact 7.2.3 guarantees that for any two different moments x, y, if $x \leq y$, then $\neg x \leq y$. Accordingly, if a certain moment is earlier than another, different one, then it is not later. This represents the common intuitions of unidirectionality of the passage of time.

The transitivity of relation $\leq$ is ensured by the ordered nature of time. Moment a that is earlier than another moment b, is also earlier than moment c in relation to which b is earlier.

Considering some moment, e.g. a, it can be stated about all moments t_1 such as $t_1 \leq a$ and $a \neq t_1$ that they are *past moments in relation to* a, and their set is *a past of moment a*. Analogically, all moments t_2 such that $a \leq t_2$ and $a \neq t_2$ we shall call *the future moments in relation to a*, whereas their set *a future of moment a*.

Conditions $i)$ and $ii)$ from the definition of time are to exclude the non-intuitive situations. One of the types of the non-intuitive situations is time branching in the past. Regardless of the way of the time perception, it always has one past. For a model with two different pasts would have led to informational paradoxes of various sorts. The second type of the non-intuitive situation involves the fact that two different moments do not share even a fragment of the common past. In such a situation, we would have been dealing with a form of a *time rupture*, an arising of the second, disjoint time structure.

The functioning of conditions $i)$ and $ii)$ on the basis of the definition of time shall be illustrated with graphs. Let the lines represent such subsets T which relation T is total. Let us assume that time *flows* from left to right. Condition $i)$ excludes the following situation.

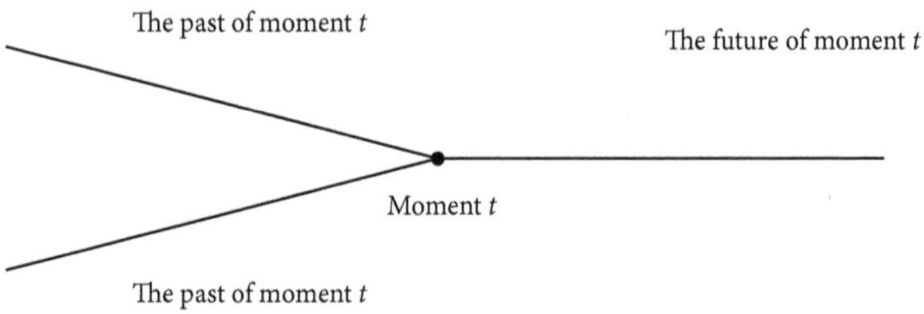

It precludes some moment of time from having two different futures. Thus, each moment has at most one future. It is not though excluded that it has many alternative futures.

However, it is possible for condition $i)$ to be satisfied, and yet the matter involving not one, but many time structures that constitute substructures of the initial time structure. Only condition $ii)$ does not allow such situation.

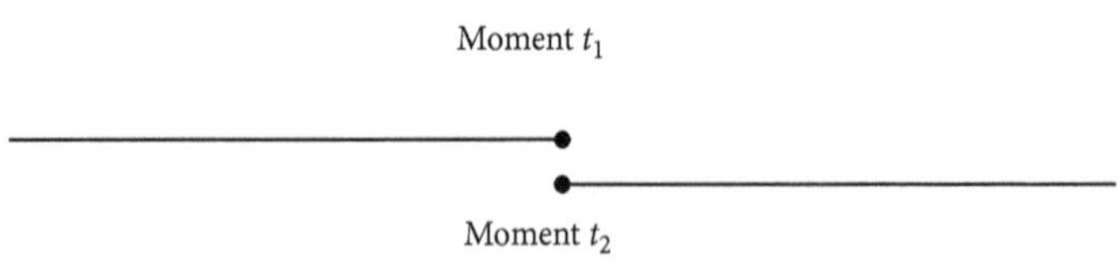

Condition $ii)$ thus excludes many such structures which interpreted as time structures *de facto* would have negated our intuitions that time is one, allowing structures, e.g. with such relation $\leq = \{\langle a,b\rangle, \langle c,d\rangle\} \cup I$. Within this structure, point a is earlier than point b, but none of the elements of this subset remains in a temporal relation with the remaining points. The intuition suggests that the matter here involves rather two different independent time structures than one. For example, moments b, d are not comparable to any common earlier point.

Properly defining the additional qualities of T and $\leq$ time of various types can be modelled. The different types of time are simply different logical structures of the time. Sometimes also the concept of a substructure of time $\langle T, < \rangle$, that is derivative to the structure of $\langle T, \leq \rangle$ due to the connection between relations $\leq$ and $<$, may be useful.

Let us now define the concept that shall allow a precise discussion about the different complete fragments of time. I mean herein the definition of the concept of the branching time.

Definition 7.2.12 *The branch of time structure $\langle T, \leq \rangle$ is any nonempty subset $T_1 \subseteq T$, such as:*

i) $\forall_{t_1, t_2 \in T_1} (t_1 \leq t_2 \vee t_2 \leq t_1 \vee t_1 = t_2)$
ii) for any set $T_2 \subseteq T$ that meets condition i), if set $T_1 \subseteq T_2$, then $T_1 = T_2$.

The branch of time is though a total set that, due to the inclusion relation, is maximal. This definition results, among other things, in the following conclusions:

Corollary 7.2.13 *(i) Each branch of time is also a time structure.*
(ii) Each time structure consists of at least one branch.

By virtue of the presented definition also, the following fact occurs:

Fact 7.2.14 *Let $\langle T, \leq \rangle$ be a time structure. If $\leq \subseteq T^2$ is total, then in set T there is exactly one branch of time.*

Proof

Assume that $\leq$ is total. Notice that T meets all the conditions imposed on the branch of time: is its own nonempty subset, $\leq$ on all its elements is total, there is no such other subset T' that T is its proper subset. Ergo, according to the definition, T is a branch of time.

Assume there is another branch of time T_1. Thus, according to the definition, there is no such nonempty set T_2, that is subset T, meets condition i) (that is, a totality condition of relation $\leq$ on all the elements of T_2) and $T_1 \subseteq T_2$ and $T_1 \neq T_2$. However, the set that meets all these conditions is T, and that excludes the existence of more than one branch of time. □

Applying the above concepts, we shall now proceed to the issue mentioned in the section on the time in physics. Its presentation will allow to state something about the time structures employed herein. This is about a non-equivalence of two definitions of *being older* in the supra described so-called father-son paradox. We shall demonstrate that the proposed concept of time guarantees the equivalence of these definitions. It will have resulted from employing the concept of the absolute time.

For the benefit of comparing the distances between the points of time, the pairs of the points should be assigned the numbers that correspond to the distances between them. The concept of the metric space will come in handy.

Definition 7.2.15 *Let T be a nonempty set of moments and $\mathbb{R}^+$ be set of positive real numbers. Function $f: T^2 \longrightarrow \mathbb{R}^+$, attributing to any two points t_i, t_j a real number called its distance, we shall name metrics on T iff this function meets the following conditions:*

i) $f(t_i,t_j) = 0 \Leftrightarrow t_i = t_j$ *(so-called an identity condition)*
ii) $f(t_i,t_j) = f(t_j,t_i)$ *(so-called a symmetry condition)*
iii) $f(t_i,t_j) + f(t_j,t_k) \geq f(t_i,t_k)$ *(so-called a triangle condition).*

A pair $\langle T, f\rangle$ we will call a metric space. Whereas the function that meets the above conditions will be said to be metrising set T.

A metric function on set T can be defined by means of function $g: T \longrightarrow \mathbb{R}^+$, defined by the condition: for any t_i, $t_j \in T$ if $t_i > t_j$, then $g(t_i) >_{\mathbb{R}^+} g(t_j)$.

Fact 7.2.16 *Let $\langle T, \leq\rangle$ be a linear order. Then T is a metric space with metric $f(t_1,t_2) = |g(t_1) - g(t_2)|$.*

The metric defined with function g can be called *a monotonically metrising function*, and the obtained metric — *a monotonic metric*. There arises the following lemma:

Lemma 7.2.17 *Let $\langle T, R\rangle$ be a linearly or strictly linearly ordered set. Then for any function f monotonically metrising set T it is so that:*

$$\forall_{t_i,t_j,t_k \in T}[t_i < t_j \wedge t_j < t_k \rightarrow f(t_i,t_k) >_{\mathbb{R}^+} f(t_j,t_k)].$$

We shall now present both definitions of *being older* on the basis of the strict order relation. Rather than stating that *x is older than y*, we shall employ a more general expression *x occurs longer than y at point t_i* and, additionally, we shall introduce an intuitive term *x begins lasting at t_j*. The definitions of *being older*, in the adopted style, will take on the following forms:

Definition 7.2.18 *Let $\langle T, \leq\rangle$ be a time structure. We shall state that x lasts longer than y at point t_i iff $\exists_{t_j,t_k < t_i \in T}$, that x begins lasting at t_j, y begins lasting at t_k, and $t_j < t_k$.*

Definition 7.2.19 *Let $\langle T, \leq \rangle$ be a time structure. Let $f : T \longrightarrow \mathbb{R}^+$ be a monotonically metrising function T. We will state that x lasts longer than y at point t_i if $\exists_{t_j, t_k < t_i \in T}$ that x begins lasting at t_j, y begins lasting at t_k and $f(t_j, t_i) > f(t_k, t_i)$.*

The concept of *lasting longer than* defined in the both ways, on the basis of our structures, designates exactly the same pairs of objects:

Theorem 7.2.20 *Let $\langle T, R \rangle$ be a strictly linear or linear order. Let $f: T \longrightarrow \mathbb{R}^+$ be a monotonically metrising function T. For any x, y: x lasts longer than y according to the def. 7.2.18 iff x lasts longer than y according to the def. 7.2.19.*

The above findings result in the following conclusions.

Corollary 7.2.21 *Each branch of time $\langle T_1, \leq \rangle$, in a monotonic metric, is a metric space.*

Corollary 7.2.22 *Within each branch of time $\langle T_1, \leq \rangle$, in the monotonic metric, the definitions 7.2.18 and 7.2.19 are equivalent.*

We consider the time and the branches of time in an absolute sense. Thus the concepts of *being earlier*, *being later* and *being simultaneous*, within the described time structures, are also considered in an absolute way.

In the further part of the study, in the subchapter on the axioms on the different types of the time structures, we shall introduce new concepts that will enable arithmetic operations on the numbers assigned to the points of time. Nevertheless, it requires a certain discernment of the branches of the time structures due to the cardinality of set T. In due course, these concepts shall be employed in the description of the functors that require arithmetic operations on the numbers that represent the moments of time.

7.2.1 The similarity of the moments, branches and time

The introduced concept of time has entirely formal nature and there are many structures corresponding to it that can be compared in terms of the formal similarity. It is being done by means of the isomorphism of the structures.

Definition 7.2.23 *Two structures of time $\langle T, R \rangle$ and $\langle T', R' \rangle$ are isomorphic (for short: $\langle T, R \rangle \simeq \langle T', R' \rangle$) iff there exists such function $f : T \longrightarrow T'$ that meets the following conditions:*

i) f is a bijection, i.e. for any $t_i, t_j \in T$:

- *if $t_i \neq t_j$, then $f(t_i) \neq f(t_j)$*
- *for any $t_i \in T'$ there is such $t_j \in T$ that $f(t_j) = t_i$*

ii) for any $t_i, t_j \in T$, $t_i R t_j$ *iff* $f(t_i) R' f(t_j)$.

Function $f: T \longrightarrow T'$ *is called an isomorphism.*

By virtue of the above definition and the characteristics of the linear structures (also the strict ones) the following fact follows.

Fact 7.2.24 *Let both time structures* $\langle T, R \rangle$ *and* $\langle T', R' \rangle$ *be the linear or strictly linear structures. For* $\langle T, R \rangle \simeq \langle T', R' \rangle$ *it would suffice if bijection* $f: T \longrightarrow T'$ *satisfied the one-sided implication from condition* ii*), i.e. one of the two conditions:*

1. *for any* $t_i, t_j \in T$, *if* $t_i R t_j$, *then* $f(t_i) R' f(t_j)$
2. *for any* $t_i, t_j \in T$, *if* $f(t_i) R' f(t_j)$, *then* $t_i R t_j$.

The definition of the isomorphic structures produces one more, useful thereafter, fact stating that the isomorphic relation defined on the Cartesian product of the set of the structures is transitive.

Fact 7.2.25 *If* $\langle X, Q \rangle \simeq \langle Y, R \rangle$ *and* $\langle Y, R \rangle \simeq \langle Z, S \rangle$, *then* $\langle X, Q \rangle \simeq \langle Z, S \rangle$.

However, in addition to the formal similarity, one can also indicate certain aspects that could be common for different temporal objects. The isomorphism of the time structures does not reveal anything about what takes place at the given time.

Certain states of affairs occur at moments. Thus, the similarity of two moments could be defined by a fact that at them exactly the same states of affairs occur[201]. However, instead of engaging in the ontological issues, the similarity of objects can be defined by referring to the definition of truthness.

Any sentence from set $\mathbf{S_{TU}}$ can be supplemented with a time parameter corresponding to the given moment, thereby obtaining a sentence from set $\mathbf{S_{TD}}$, that is associated exclusively to this moment. This is the way to define the relation of similarity between moments.

Definition 7.2.26 *About any two points of time* t_i, t_j *it can be said that they are similar (for short:* $t_i \approx_m t_j$*) iff they meet the following condition:*

for any sentence $\mathbf{S_{TU}}$, *sentence* s *is true at* t_i *iff it is true at* t_j.

---

201 Intuitively, we could assume that the states of affairs are closed against considering them in accordance to the moments of their occurrence. For example, if a state of affairs takes place at $n > 1$ moments, then its part which occurs, say, at $n - 1$ moments is also a state of affairs.

Two moments that are similar $t_i \approx_m t_j$, are similar in relation to set of sentences $\mathbf{S}_{\mathrm{TU}}$, therefore in relation to the states of affairs that in general are linguistically describable. Nevertheless, what cannot be described shall not be discussed herein. Even if two moments differ in a state of affairs that cannot be described, it will not affect logical values of any simple sentences.

Fact 7.2.27 *Relation $\approx_m$ is an equivalence relation, i.e. it meets the conditions of: reflexivity, symmetry and transitivity.*

Notice that between the identity relation of moments and similarity relation defined on a set of moments, there occurs the following relationship.

Corollary 7.2.28 *For any two moments t_i, t_j, if $t_i = t_j$, then $t_i \approx_m t_j$.*

The reference to the concept of the sentences undetermined in time within the definition of similarity has been intentional, because the use of the concept of the sentences determined in time would have subjected the similarity of two objects from the question of how things are at the other, earlier or later moments. For the sentences determined in time can just as well state about the future as the past moments, depending on the considered points of time. Whereas, the similarity of the points should only consider what takes place within their range. However, since the sentences undetermined in time, supplemented with a time parameter corresponding to the considered points, become sentences determined in time, therefore apparently the following conclusion occurs.

Corollary 7.2.29 *For any moments t_i, t_j, if for any sentence $\mathbf{S}_{\mathrm{TD}}$, sentence s is true at moment t_i if and only if it is true at moment t_j, then $t_i \approx_m t_j$.*

Employing the concept of similarity of moments we also want to define the concept of material similarity of time structures.

Definition 7.2.30 *Two time structures $\langle T, R \rangle$ and $\langle T', R' \rangle$ are materially similar iff there exists such bijection $f : T \longrightarrow T'$ that for any $t_i, t_j \in T$:*

i) $t_i R t_j$ iff $f(t_i) R' f(t_j)$ (hence $\langle T, R \rangle \backsimeq \langle T', R' \rangle$)
ii) $t_i \approx_m f(t_i)$.

One more notion shall be applicable for the analysis of the issue of determinism. For, in spite of the same content of events, due to an issue of the logical values, two scenarios of the history of the world could have varied. For example, the same states of affairs could belong to the two different histories of the world, but in a different order. Even though the sentences undetermined in time supplemented with the proper values of the time parameter would have been then stating about different moments within the given structures, yet they would have been described by exactly the same simple sentences. Such time structures will be called the weakly materially similar structures.

Definition 7.2.31 *Two time structures $\langle T, R\rangle$ and $\langle T', R'\rangle$ are weakly materially similar iff they meet the following condition:*

$$\forall_{z \in \mathbf{S}_{\mathrm{TU}}} (\exists_{t \in T}\, z \text{ is true at } t \Leftrightarrow \exists_{t' \in T'}\, z \text{ is true at } t').$$

Both the above definitions are of course applicable to the branches of time which are a special case of the time structures. Between the concepts of the materially similar structures and weakly materially similar ones, there occurs the following relation:

Corollary 7.2.32 *If any two time structures $\langle T, R\rangle$ and $\langle T', R'\rangle$ are materially similar, they are also weakly materially similar.*

Notice that, in relation to the formal scientific research, the material similarity of structures is an intangible notion. Therefore, the minimal condition of the similarity of structures is their formal similarity. After this general conceptual introduction, let us proceed to the notions that will allow to narrow down the overall concept of time.

7.3 Axioms of the logical structures of time

The different concepts considering time, which we have introduced in the subchapter on the formal representation of the time, are of the general form. It considers the very concept of time in particular. Now, we shall present the axioms that will allow time modelling just through narrowing down the general concept of time by means of imposing additional conditions on it.

One of the basic qualities of the time structures is including a moment that is first or last within the structure. Intuitively, this is about the model of time with a beginning or an end.

Definition 7.3.1 *Let $\langle T, \leq\rangle$ be a time structure. Structure $\langle T, \leq\rangle$ has a beginning iff it meets the following condition:*

$$\exists_{t_i \in T} \forall_{t_j \in T} (t_i \leq t_j).$$

Definition 7.3.2 *Let $\langle T, \leq\rangle$ be a time structure. Structure $\langle T, \leq\rangle$ has an end iff it meets the following condition:*

$$\exists_{t_i \in T} \forall_{t_j \in T} (t_j \leq t_i).$$

From the above definitions and the definitions of time and the branches, the following conclusion follows.

Corollary 7.3.3 *Let $\langle T, \leq \rangle$ be a structure of time that has a beginning (respectively an end). Then any branch of this structure has a beginning (respectively an end).*

Of course, in the case of the structure with an end, the opposite conclusion is not true. It is possible that each branch as a time structure has an end, but the structure containing it does not have an end.

The structure that has a beginning and an end will meet the conditions from both the aforementioned definitions.

Another important quality of the time structures is associated with the question whether between any two moments, earlier and later ones, there is always some medium moment, later than the first and earlier than the second one. One of the answers to this question is a notion of dense time.

Definition 7.3.4 *Let $\langle T, \leq \rangle$ be a time structure. Structure $\langle T, \leq \rangle$ is dense iff it meets the following condition:*

$$\forall_{t_i, t_j \in T} (t_i \leq t_j \rightarrow \exists_{t_k \in T} (t_k \neq t_i \wedge t_k \neq t_j \wedge t_i \leq t_k \wedge t_k \leq t_j)).$$

However, we can consider such model of time where each moment is preceding to some moment without another intermediate moment, and subsequent to some moment without another intermediate moment. By this we mean a notion of discrete time. Discrete structure is a structure where for each element (alternatively excluding its end), there is an element directly subsequent (then we call it a right-discrete structure) and for each element (alternatively excluding its beginning), there is an element directly precedent (then we call it a left-discrete structure). In the formal description it is easier, though due to the fact of 7.2.9 equally, to apply relation <:

Definition 7.3.5 *Let $\langle T, \leq \rangle$ be a time structure. Structure $\langle T, \leq \rangle$ is discrete iff meets the following conditions:*

(Right discreteness):

$$\forall_{t_i, t_j \in T} (t_i < t_j \rightarrow \exists_{t_k \in T} (t_i < t_k \wedge \neg \exists_{t_l \in T} (t_i < t_l \wedge t_l < t_k))$$

(Left discreteness):

$$\forall_{t_i, t_j \in T} (t_i < t_j \rightarrow \exists_{t_k \in T} (t_k < t_j \wedge \neg \exists_{t_l \in T} (t_k < t_l \wedge t_l < t_j))).$$

The consideration of the time structures, of which some subsets are dense and others discrete, that is, of the mixed time structures, is not excluded. Nonetheless, we shall not be considering such structures. The discussed structures will be either dense or discrete. In particular, we shall work on such dense structures that are continuous. The notion of continuity shall be introduced in due course.

Apart from the linear structures, also non-linear ones can be considered. These are interesting in the light of the problem of the determinism of the sentences about the future. Such structures shall be called branching structures, since they have more than one branch of time. Within these structures, there can be found three different points, of which the first one precedes the other two that are not mutually comparable. Formally:

Definition 7.3.6 *Let $\langle T, \leq \rangle$ be a time structure. Structure $\langle T, \leq \rangle$ is branching (branching into the future) iff it meets the following condition:*

$$\exists_{t_i, t_j, t_k \in T} (t_i \leq t_j \wedge t_i \leq t_k \wedge \neg t_j \leq t_k \wedge \neg t_k \leq t_j).$$

A set of the points that are later or identical to t_j and, respectively, a set of the points that are later or identical to t_k, is called branching.

From the above definition and from the definition of the branch of time, another conclusion follows.

Corollary 7.3.7 *Let $\langle T, \leq \rangle$ be a time structure branching into the future. Then $\langle T, \leq \rangle$ has at least two different branches of time.*

A similar measure can be applied in defining the structure branching into the past in relation to a certain point, whereas, putting this definition together with the condition from the above definition, we could have obtained a definition of a structure that is branching into the future as well as into the past. Such time structures are though excluded by virtue of the definition of the time structure; for each point has unambiguously determined past within the structure.

For the further procedures regarding introducing the logical tools will also prove useful, apart from the time structures, the notions that would allow performing arithmetic operations on the numbers assigned to the moments of time. Coordinating is herein a basic notion. Prior to expressing the definition of coordinating, we shall specify that the dense structures will be coordinated by real numbers, whereas the discrete structures — by integers.

In order to develop further concepts, a reminder of certain issues concerning the concept of continuity, e.g. by the Dedekind cut, is necessary. Take any nonempty linearly ordered set $\langle X, R\rangle$. Cut is such a pair $\langle X_1, X_2\rangle$ of subsets X that meets the three following conditions:

$i)$ $X_1 \cup X_2 = X$,
$ii)$ $X_1 \cap X_2 = \varnothing$,
$iii)$ $\forall_{x \in X_1} \forall_{y \in X_2} xRy$.

We state that subset X_1 *has a last element* iff the following condition is met:

$$\exists_{x \in X_1} \forall_{y \in X_1} yRx.$$

Whereas, we shall state that subset X_2 *has a first element* iff the following condition is met:

$$\exists_{x \in X_2} \forall_{y \in X_2} xRy.$$

If in the cut $\langle X_1, X_2\rangle$ set X_1 does not have the last element and set X_2 does not have the first element, then this cut *yields a gap*. Linearly ordered set $\langle X, R\rangle$ will be called *continuous* iff it is dense and none of its cuts $\langle X_1, X_2\rangle$ yields a gap.

The dense linear time structures included in the procedure of coordinating the numbers will not have gaps within any of their cuts. Thus, they will be the linearly ordered sets of continuously ordered type. While the dense structures, which we will not coordinate, we do not assume to be continuous.

An example of a set that is dense, but not continuous is set $\mathbb{R} \setminus \{0\}$. Within this set, there can be indicated such cut $\langle \mathbb{R}_1, \mathbb{R}_2\rangle$ that $\mathbb{R}_1 = \{x \in \mathbb{R} : 0 > x\}$ and $\mathbb{R}_2 = \{x \in \mathbb{R} : x > 0\}$. The first set does not have the biggest element (for the numbers there increase to zero), as well as the second one does not have the smallest element (for the numbers there decrease to zero).

Apart from the concept of the continuity, there will be also necessary one more concept associated with the numbers. Let $\langle X, R\rangle$ be a set of numbers, linearly ordered by relation R (natural numbers, integers, rational numbers or real numbers). Take its any subset $Y \subseteq X$. We shall state that Y is *complete* iff

$$\forall_{x, y \in Y} (xRy \rightarrow \forall_{z \in X} (xRz \wedge zRy \rightarrow z \in Y)).$$

An example of the linearly ordered set of the integers, that, according to the above concept, is not complete, is set $\mathbb{Z}' = \{ -5, 0, 1, 2 \}$, for, among others, it does not include the consequent of number -5, that is, number -4.

By means of the notions of the continuous set and the complete set, we shall now define concepts of a proper coordination and of a coordination. Let us also add that if set X' is a subset of some set $\langle X, R\rangle$, then the notation: $\langle X', R|X'\rangle$ denotes the structure in which the initial relation R has been cut down to set X'.

Definition 7.3.8 *Let $\langle T, R\rangle$ be a time structure. We shall state that linearly ordered set of numbers $\langle X, \leq'\rangle$ well coordinates set $\langle T, R\rangle$ iff:*

i) $\langle X, \leq'\rangle$ is complete
ii) $\langle T, R\rangle \simeq \langle X, \leq'\rangle$.

A time structure that has a well-coordinating structure will be called a well-coordinated structure; the isomorphism that assigns a number to each moment will be called a coordinating function and denoted as f^k; and a structure well coordinating time structure $\langle T, R\rangle$ we will be denoting as $\langle T^k, R^k\rangle$.

The definition 7.3.8 states that well coordinated can be, at the very most, certain linear time structures, and particularly, the branches of time. However, since the considered herein time structures may be branching into the future, we shall introduce an additional concept that will allow to assign numbers to all points of time, also within the branching structures.

Definition 7.3.9 *Let $\langle T, R\rangle$ be a time structure. Structure $\langle T, R\rangle$ will be called coordinated iff there exists the smallest set of coordinating functions F such that:*

1. *for each branch $T' \subseteq T$ there is exactly one coordinating function $f'^k \in F$*
2. *for any moment t_i and any branches T', $T'' \subseteq T$ if $t_i \in T' \cap T''$ and $\{f'^k, f''^k\} \subseteq F$, then $f'^k(t_i) = f''^k(t_i)$.*

A set of the functions that coordinate given time structure $\langle T, R\rangle$ will be denoted with a symbol F, optionally with an index.

Since the following conclusion occurs:

Corollary 7.3.10 *Each well-coordinated structure is coordinated.*

then, if there is no need to indicate the difference, we shall simply call them the coordinated structures.

From the above definition also, another conclusion follows:

Corollary 7.3.11 *Each discrete linear time structure has a well-coordinating structure.*

Summarising this section, we shall draw attention to several issues. In the first place, while employing the coordinating structures, we shall use the linear structures of numbers, in which the numbers are ordered increasingly. Thus, the numbers assigned to the future moments will be larger than the numbers assigned to the past moments.

The coordinating structures allow to treat the time in a dynamic manner. Within the given time structure, there can be distinguished some moment n called a present moment, and within the coordinating structure assigned number 0. Thus, the preceding moments would have had negative numbers assigned, and the subsequent moments — positive numbers. Employing the fact 7.2.25 stating about the transitivity of isomorphism, the present moment can be moved by adding to the initial coordinating structure, or subtracting from it, a certain number, it resulting in another coordinating structure, in which the status of the present, that is, number 0, corresponds to another moment.

According to the Definition 7.3.9 within the branching time structures, the moments from the branches are unambiguously assigned numbers, but not necessarily in a single unambiguous way. Take e.g. the following substructure of any branching structure $\langle\{a,b,c,d\}, \{\langle a,b\rangle, \langle b,c\rangle, \langle b,d\rangle, \langle a,c\rangle, \langle a,d\rangle\}\rangle$. Assuming that $n = b$, the individual branches can be well coordinated e.g. in a following manner: $\langle a,b,c\rangle$ is well coordinated by $\langle -1,0,1\rangle$, while $\langle a,b,d\rangle$ is well coordinated by $\langle -1,0,1\rangle$. Since each branch is well coordinated and to the corresponding within the branches moments the same numbers are assigned, the branching time structure is coordinated.

7.4 The outline of logics of time $\mathcal{R}_n^+$

We shall outline the logics of coordinated time, i.e. the time in which numbers are assigned to the individual moments. By means of this operation, it is possible — by performing the relevant arithmetic operations — to *move* in time, forwards and backwards. From the other systems of logics with operator $\mathcal{R}_i$, they differ, inter alia, in the fact that in each time structure there can be distinguished a certain point called the present (denoted by n)[202].

Nonetheless, the most important difference, in terms of operator $\mathcal{R}_i$, will concern its interpretation. Within the logics developed, among others by Rescher, it is interpreted in a direct way. The interpretation is a certain ordered pair $\langle T,F\rangle$, where T is a time structure and F is a set of the functions that assign logical values to each sentential variable, in the given point of time. Simplified, formula $\mathcal{R}_a(p)$ is true if and only if in the point denoted by a variable p takes a value of one.

This point of view can be, however, modified by interpreting the operator in the following way. Formula $\mathcal{R}_a(p)$ is fulfilled at point a if and only if in some branch of time a there is such point of time b that variable p takes the

202 For a change, Rescher and Urquhart describe the logics with operator $\mathcal{R}_i$, where n acts as an occasional expression, not denoting any specific point of time [241].

value one. This idea has been based on the following intention. The sentences determined in time state something about certain determined moments in time in which the states of affairs, predicated in the sentences, occur. Therefore, within these moments of time, the sentences are either true or false, whereas in the subsequent or preceding moments, they are being realised (assuming partially-ordered time structures will have guaranteed that the sentences true in the given moment are also realised within it). Thus, in this interpretation, the realisability will be a weaker concept than the truthness. Within the developed herein concept of the logics of the coordinated time, we shall indicate some axioms that would have corresponded to the realisability.

Both the issues result in the notions required for accurately describing such approach as being quite complicated, thus the outline that shall be presented herein requires further work. In particular, it relates to the axioms with a new understanding of operator $\mathcal{R}_i$.

First, we shall describe the grammar of logic $\mathcal{R}_n^+$ along with its specific axioms, and afterwards we shall characterise the semantic concepts.

7.4.1 Grammar

The language will be described by means of metavariables. Also useful will be the metavariables taking values from a set of formulas: A, B, C, ... and the metavariables taking values from a set of indices: i, j, k, ... The definition of the formula and the index shall be presented in due course.

Since logics $\mathcal{R}_n^+$ are the structural extensions of **CPL**, then in their language there are surely all the extensional, one-argument and two-argument, functors. Moreover, a modality with an index and quantifiers, used to quantifying the variables present in the indices, should be emphasised. Thus the set of the logical constants is of the following form: $SL = \{\neg, \wedge, \vee, \rightarrow, \leftrightarrow \mathcal{R}_i, \forall, \exists\}$.

Some points of time and distances between the points can be denoted by so-called index constants. There are the following objects of this type in the alphabet: $Sc = \{n, a, b, c, a_1, b_1, c_1, \ldots\}$ for the point constants $So = \{f, g, h, f_1, g_1, h_1, \ldots\}$ for the distance constants.

Within the language, there are also identified sentential variables: $Sv = \{p, q, r, \ldots, p_1, q_1, r_1, \ldots\}$, variables for the points of time: $Vt = \{t_1, t_2, \ldots\}$ and variables for distances: $Vd = \{d_1, d_2, \ldots\}$.

In addition to the above symbols, also operational constants should be distinguished. A set of the operational constants consists of two functional symbols: $Cop = \{-, +\}$.

Moreover, to preserve the unambiguity of the notation, we shall need also the auxiliary symbols: $Ac = \{), (,], [\}$.

Now, we shall present the definitions of constructing the correct expressions of the considered logic.

Definition 7.4.1 *A set of indices I is the smallest set of expressions X that meets the following conditions:*

1. $Sc \cup Zc \subseteq X$
2. $\{x : x := k + j \text{ or } x := k - j\} \subseteq X$, *where* $k \in X$ *and* $j \in So \cup Zo$.

$\mathsf{For}_{\mathrm{CPL}}$ will stand for a set of all the classical logic formulas. Let us now define the concept of formula $\mathcal{R}_n^+$:

Definition 7.4.2 *A set of the formulas* $\mathsf{For}_{\mathcal{R}_n^+}$ *is the smallest set Xthat meets the following conditions:*

1. $\mathcal{R}_i(A) \in X$, *for any* $A \in \mathsf{For}_{\mathrm{CPL}}$ *and* $i \in I$
2. $\neg A$, $(A \wedge B)$, $(A \vee B)$, $(A \rightarrow B)$, $(A \leftrightarrow B)$ *belong to X if A, B belong to X*
3. $\forall_i A \in X$, *for any* $A \in X$ *and* $i \in Zc \cup Zo$
4. $\exists_i A \in X$, *for any* $A \in X$ *and* $i \in Zc \cup Zo$.

The elements of set $\mathsf{For}_{\mathcal{R}_n^+}$ *are called the formulas.*

7.4.2 Axioms and rules of deduction

Apart from the rules and axioms characteristic for the classical logic, there are also specific axioms in the described logic. Of course, we shall be providing the axiom schemata, albeit putting it simply as the axioms.

The basic axiom scheme for all the logics with operator $\mathcal{R}_i$ superstructured upon the classical logic is the following one:

$$A_{\mathcal{R}1}\ \mathcal{R}_i(A)$$

where A is an assertion in **CPL** and i is an index. Thus, each assertion in the classical logic is realised at any point of the time structure. Notice that axiom $A_{\mathcal{R}1}$ subsequent to the acceptance of the idea of realisability, 'imposes' the following quality on the time structure: from each point in the time structure, some point of time must be accessible — earlier, later or the one. This axiom, along with the following two:

$$A_{\mathcal{R}2}\ \mathcal{R}_i(\neg A) \leftrightarrow \neg\mathcal{R}_i(A)$$

$$A_{\mathcal{R}3}\ \mathcal{R}_i(A) \wedge \mathcal{R}_i(B) \rightarrow \mathcal{R}_i(A \wedge B)$$

allows to prove the distribution of operator $\mathcal{R}_i$ across all the functors of the classical logic[203].

Another two axioms characterise the discussed time structures. Going back in time, we do not loose the possibilities of the realisation of the sentence, but with the passage of time, we may.

$$A_{\mathcal{R}4}\ \mathcal{R}_k(A) \to \mathcal{R}_{k-j}(A)$$
$$A_{\mathcal{R}5}\ \mathcal{R}_{k+j}(A) \to \mathcal{R}_k(A).$$

Another schema corresponds to the arithmetic character of the operations carried out on the moments of time:

$A_{\mathcal{R}6}\ \mathcal{R}_i(A) \to \mathcal{R}_j(A)$, as long as equality $i = j$ is a law of the arithmetics of real numbers.

Example. The following formula $(t + d) - d) = t$ is a law of arithmetics of real numbers after interpreting variables t, d as the variables ranging over the domain of real numbers. Ergo $i = (t + d) - d) = t = j$ and in consequence formula: $\mathcal{R}_{(t+d)-d}(p) \to \mathcal{R}_t(p)$ is a special case of the schema $A_{\mathcal{R}6}$.

Assume that notations $A(i), A(j)$ state that in formula A in the first case there occurs index i and in the second case instead of index i exactly in the same places within formula A occurs index j and it does not occur in any other places within formula A. Naturally, also the reverse situation occurs. Thus, the two formulas differ structurally only in the fact that each occurrence of index i have been replaced with the occurrences of index j, and vice versa. In addition to the above axioms, we propound another one that extends the described logic and holds over only in certain special time structures:

$$A_{\mathcal{R}7}\ \exists_{t_1} \forall_{d_1} (A(t_1 - d_1) \wedge A(t_1) \wedge A(t_1 + d_1)) \to \forall_{t_2} A(t_2).$$

This axiom could be weaker, that is, in the antecedent of the implication there could have been e.g. no module $A(t_1 - d_1)$. For it results from axiom $A_{\mathcal{R}4}$ and from the laws of the classical logic $A(t_1)$. However, it will prove useful in its stronger variation without the assumption of axioms $A_{\mathcal{R}4}, A_{\mathcal{R}5}$.

Axiom $A_{\mathcal{R}7}$ states that if there exists such point of time that at it, as well as at each future and each past ones, sentence A is true, then at each point it is true.

Of course, within the given language, we can also present other interesting axioms. The remaining syntactic notions are standard.

---

203 See [119]. Of course, also the rule of detachment should be adopted.

7.4.3 Semantics

The formal description of the semantics for the presented systems has to be quite complicated, for there are many symbols in the alphabet. Therefore, we shall restrict this analysis to certain informal remarks.

The time structures, within which we interpret the formulas of logics $\mathcal{R}_n^+$, must allow to be well coordinated. Thus, with each structure, there should be associated at least one set F of all functions f^k that coordinate individual branches of time within this time structure, within the determined coordinating structure. Moreover, the coordinating structure assigns number zero to exactly one moment. In the given interpretation, this very moment is considered present. If, for the given structure, there exist more sets F attributing number zero to the number of the moments of time different than one, then it means that the time is not static. A change of coordinating structure conveys a shift of time, its passage.

In order to interpret the arithmetic signs as the additive and subtractive operations carried out on the values assigned to the moments, the coordinating structures have to be closed under these operations. It means that their domains must be the sets of numbers in which these operations can be carried out.

If e.g. structure is discrete, then the domain of its coordinating structure is set of integers $\mathbb{Z}$. If the structure is continuous, then the domain of its coordinating structure is set of real numbers $\mathbb{R}$. Therefore, within the given time structure, all the branches should be well coordinated in the linearly ordered set of the integers or real numbers. The important issue is that the execution of arithmetic operations requires the branches neither to have an end nor a beginning. Then, starting at a given moment denoted by t_1 it is possible to determine the moment of the branch that is denoted by symbol $t_1 + d$ (or e.g. $t_1 - d$), for any number l from the domain of the coordinating structure.

Beyond doubt, such conditions are met by the linear time structures with no end and no beginning, that are coordinateable by such sets F in which the coordinating functions allow to perform arithmetic operations (actually, in each F there is exactly one function). Within these structures, moment n can also be moved freely into the past or future.

Another type of the structures that are suitable for it, are some structures of the branching time in which the branches also neither have an end nor a beginning. But in their case, moment n — coordinated by number zero — cannot be freely moved, as its shift into the branch of time that has an alternative branch would have meant that there exists another one, parallel present point of a coordinate equal to zero. However, in principle, within each coordinating structure there is to be only one present moment that in the language corresponds to symbol n. Thus, with such an approach to the present point n the time structures appropriate for the discussed logics must have such quality that for any preceding moment n the branching cannot occur. Admittedly, it is theoretically possible to consider the branching in moments preceding n, but only such that create branches shorter than the amount of the moments between the branching and moment n. Then, however, these branches would have had an end, and this would exclude the execution of some arithmetic operations.

Naturally, the existence of the greater amount of the constants for the alternative presents could be allowed. However, deciding to singularise within an alphabet one constant n with a denotation in the point zero, we emphasise the fact of a loss of the other possibilities during the passage of time. The problem described herein presents, to some extent, that with these assumptions (a single present zero in each interpretation and the executability of the arithmetic operations) the forward shift in time enforces exclusion and does not allow to consider, nor even view as alternative, the branches of time that have detached from the main time axis in the past. It gives rise to an idea to represent the passage of time in a more dynamic manner, rejecting all the branches of the past on every shift of the present. In practice, this would mean that the forward shift in time by m moments involves a reduction of the time structure to its substructure in which all the alternative branches, that we have passed by on the way, are being rejected. On the one hand, this is another idea of the description of logics $\mathcal{R}_n^+$, on the other hand, though, it is a new possibility of formal representation of the passage of time — by reducing the time structure of the branches from the past. However, none of these ideas will be further developed within this study.

It is clear that discussing the interpretations of the phrases from language $\mathcal{R}_n^+$ requires introducing many additional ideas concerning the understanding of the formulas. We shall carry it out concisely by indicating the basic elements which do not require decision-making. While describing the semantics for logics $\mathcal{R}_n^+$ the functions attributing moments to the temporal constants and numbers to the distance constants should be taken into account. Moreover, the functions attributing numbers to the temporal and distance variables will also be needed. On the set of the moments and classically structured sentences, there are defined also functions $V : T \times \mathbf{For}_{\mathrm{CL}} \longrightarrow \{0,1\}$ which in each moment of time attribute logical values to each sentence, providing for the boolean structure of the sentence.

In the case of each expression $\mathcal{R}_i(A)$ with constants in index i, there should exist a possibility to unambiguously determine the moment (or a set of the moments — in a more complicated interpretation) that denotes index i — for short $d(i)$. According to the weaker interpretation of the realisability, in the simplest approach sentence $\mathcal{R}_i(A)$ would have been true if and only if within the time structure there exists moment t such as $t \leq d(i)$ or $d(i) \leq t$ and $V(t, A) = 1$. The other expressions structured from quantifiers and functors are interpreted standardly.

In the view of the philosophical nature of this book, we are omitting the technical issues and do not address the considerations about the relationship between axiomatics and semantics. We shall only mention that in such constructed semantics all the given axioms — from $A_{\mathcal{R}1}$ to $A_{\mathcal{R}6}$ — are true within any time structures congruent with logics $\mathcal{R}_n^+$, whereas axiom $A_{\mathcal{R}7}$ distinguishes the structures that are linear. It can be confirmed by considering sentence p and branching structure within which sentence p will be assigned a value one in just one moment t_1 on one of the branches; in the remaining points it will be false. Of course there will be such point — point t_1 that for any distance d the following formulas are true:

(a) $\mathcal{R}_{t_1 - d}(p)$
(b) $\mathcal{R}_{t_1}(p)$
(c) $\mathcal{R}_{t_1 + d}(p)$.

However, it does not mean that for any t_2 formula, $\mathcal{R}_{t_2}(p)$ is true, for on the other branch in none of the points available after the branching, sentence p is true.

7.5 Tense logic

The tense logic is a logic of the sentences undetermined in time $\mathbf{S}_{\mathbf{TU}}$. As mentioned above, the basic grammatical tenses are represented by means of special operators. The sentence stating about the sea battle can be uttered meaning various ways of the realisation of such state of affairs as a sea battle, e.g. as follows:

1) *It has been the case that there is a sea battle.*
2) *It has always been the case that there is a sea battle.*
3) *There is a sea battle.*
4) *It will always be the case that there is a sea battle.*
5) *It will be the case that there is a sea battle.*

Word ‘is’ within the individual sentences obviously means that the sea battle is occurring at a certain point of time in relation to the determined one, sequentially: at least in one past, in each past, in the present one, in each future and at least in one future moment. The tense logic has the means to express each of the given possibilities. Respectively the individual types of the sentences can be represented by means of the following schemata: 1) $\mathsf{P}p$, 2) $\mathsf{H}p$, 3) p, 4) $\mathsf{G}p$, 5) $\mathsf{F}p$. Therefore, operators $\mathsf{P}, \mathsf{H}, \mathsf{G}, \mathsf{F}$ have the syntactic nature of the sentence-forming functors of a sentential argument. Usually in the language of tense logic there are introduced only H and G, the remaining two operators are defined by means of the former ones and the negation functor.

Apart from the presented tense operators, to the alphabet of the language of the tense logic belong also classical functors: $\neg$, $\wedge$, $\vee$, $\rightarrow$, $\leftrightarrow$ and, by default, the set of sentential variables **Var** that are ranging over the set of the temporally indeterminate sentences, and the brackets. Additionally, we will also employ metavariables A, B, C, …, ranging over the set of all the formulas. Prior to presenting the definitions of operators F and P, we shall introduce the formulas of the tense logic.

Definition 7.5.1 *The set of the formulas of the tense logic is the smallest set X meeting the following conditions:*

i) $\mathbf{Var} \subseteq X$
ii) if $A, B \in X$, *then* $\neg A$, $(A \wedge B)$, $(A \vee B)$, $(A \rightarrow B)$, $(A \leftrightarrow B)$, $\mathsf{H}A$, $\mathsf{G}A \in X$.

The set of all the formulas of the tense logic will be denoted with symbol For^T. *The elements of set* For^T *are called formulas.*

Functors F and P are introduced to the language by means of the following definitions.

$$DEF.^{\mathsf{F}}\ \mathsf{F}A \equiv \neg\mathsf{G}\neg A$$

$$DEF.^{\mathsf{P}}\ \mathsf{P}A \equiv \neg\mathsf{H}\neg A.$$

The definition, e.g. first, states that *it will be the case that A if and only if it is not so that it will always be the case that not A*, therefore, that once it will be the case that A. However, a good substantiation of their appropriateness requires to present the concept of the interpretation of the language formulas and to specify the conditions of the truthness for the various types of its expressions.

In the tense logic, the interpretation is an ordered triple $\langle W, R, V \rangle$. W is a nonempty set of certain objects that are usually understood as the global states of affairs, that is, the temporal parts or the points of time. They can also be seen as certain special possible worlds, i.e. different alternatives of the given world that can change over time in many ways. R is a relation defined on W^2, understood as the relation of accessibility between the elements of W. The last element of the ordered triple is a two-argument function assigning logical values to the sentential variables within the individual temporal parts. The formal interpretation of the formulas of the tense logic can be presented as follows:

Definition 7.5.2 *Interpretation is* $\mathfrak{I}^T = \langle W, R, V \rangle$, *where sequentially:*

i) W *is a nonempty set of certain elements called points*
ii) R *is a relation defined on* W: $R \subseteq W^2$
iii) V *is any function:* $V: W \times \mathbf{Var} \longrightarrow \{0,1\}$.

Function valuating variables V can be unambiguously extended to the function valuating all the formulas, by means of the recursive definition of fulfilment for various types of the expressions.

Definition 7.5.3 *Let* $\mathfrak{I}^T = \langle W, R, V \rangle$ *be an interpretation. Valuating function* For^T *is an extension* $\overline{V}$ *of function* V *to set* For^T, *that for any* $x \in W$, $p_i \in \mathbf{Var}$, $A \in \mathsf{For}^T$ *meets the following conditions:*

i) $\overline{V}(x, p_i) = V(x, p_i)$
ii) $\overline{V}(x, \neg A) = 1 \Leftrightarrow \overline{V}(x, A) = 0$
iii) $\overline{V}(x, A \wedge B) = 1 \Leftrightarrow \overline{V}(x, A) = 1 \text{ and } \overline{V}(x, B) = 1$
iv) $\overline{V}(x, A \vee B) = 1 \Leftrightarrow \overline{V}(x, A) = 1 \text{ or } \overline{V}(x, B) = 1$
v) $\overline{V}(x, A \rightarrow B) = 1 \Leftrightarrow \overline{V}(x, A) = 0 \text{ or } \overline{V}(x, B) = 1$
vi) $\overline{V}(x, A \leftrightarrow B) = 1 \Leftrightarrow \overline{V}(x, A) = \overline{V}(x, B)$
vii) $\overline{V}(x, \mathsf{G}A) = 1 \Leftrightarrow \forall_{y \in W} (xRy \Rightarrow \overline{V}(y, A) = 1)$
vi) $\overline{V}(x, \mathsf{H}A) = 1 \Leftrightarrow \forall_{y \in W} (yRx \Rightarrow \overline{V}(y, A) = 1)$.

By means of these conditions and $DEF.^{\mathsf{F}}$ and $DEF.^{\mathsf{P}}$ the following conditions for the formulas in forms $\mathsf{F}A$ and $\mathsf{P}A$ can be introduced:

Corollary 7.5.4 *Let* $\overline{V}$ *be valuating function* FOR^T. *Then for any* $x \in W$, $A \in FOR^T$ *it is so that:*

i) $\overline{V}(x, \mathsf{F}A) = 1 \Leftrightarrow \exists_{y \in W} (xRy \,\&\, \overline{V}(y, A) = 1)$
ii) $\overline{V}(x, \mathsf{P}A) = 1 \Leftrightarrow \exists_{y \in W} (yRx \,\&\, \overline{V}(y, A) = 1)$.

Having the definition of the formula and the valuation of formulas, now we can define the formula fulfilled in the given interpretation.

Definition 7.5.5 *Let* $\mathfrak{I}^T = \langle W, R, V \rangle$ *be an interpretation. Moreover, let* $A \in \mathsf{For}^T$ *and* $x \in W$. *We shall state that formula* A *is true in* $\mathfrak{I}^T$, *at point* $x \in W$ *(for short:* $\mathfrak{I}^T \vDash_x A$*) if and only if* $\overline{V}(x, A) = 1$.

By means of the above definition, the concept of the logical consequence and tautology of the tense logic can be determined.

Definition 7.5.6 *We shall state that from the set of formulas* $X \subseteq \mathsf{For}^T$ *follows formula* $A \in \mathsf{For}^T$ *(for short:* $X \models A$*) if and only if for any interpretation* $\mathfrak{I}^T = \langle W, R, V \rangle$ *and any* $x \in W$*, if for any* $B \in X$*,* $\mathfrak{I}^T \models_x B$*, than* $\mathfrak{I}^T \models_x A$*.*

Definition 7.5.7 *We shall state that formula* $A \in \mathsf{For}^T$ *is a tautology of the tense logic if and only if A follows from the empty set of formulas.*

Naturally, from the two above definitions, the following conclusion results.

Corollary 7.5.8 *Formula* $A \in \mathsf{For}^T$ *is a tautology if and only if for any interpretation* $\mathfrak{I}^T$ *and any* $x \in W$ *it is so that* $\mathfrak{I}^T \models_x A$*.*

The above concepts are defined very generally. For the interpretations can differ not only in the amounts of the elements of W and in the valuation of the sentential variables V, but also in the formal qualities of relations of accessibility R.

By imposing various conditions on R, e.g. transitivity, density etc., we get certain classes of interpretations, called *frames*. Many of these classes can be axiomatised by means of specific axioms of the tense logics, in such manner that all and only such formulas that result from them are fulfilled in the given class of interpretation. Let us now employ the review of the axiomatics of the tense logics from the work of Robert McArthur [181] and present some of the sets of the axioms characterising certain classes of interpretation, indicated by the qualities of relation R.

The basic system of tense logics is a system called $\mathbf{K}_t$. It includes all the laws of the classical propositional logic and rule of detachment MP. Moreover,
its specific axioms include the following formulas:

$$A_1^T \ \mathsf{G}(A \to B) \to (\mathsf{G}A \to \mathsf{G}B)$$

$$A_2^T \ \mathsf{H}(A \to B) \to (\mathsf{H}A \to \mathsf{H}B)$$

$$A_3^T \ A \to \mathsf{HF}A$$

$$A_4^T \ A \to \mathsf{GP}A.$$

Apart from MP, the inference rules $\mathbf{K}_t$ include also the two following rules:

R_1^T Rule of introducing G:

$$\frac{\vdash A}{\vdash \mathsf{G}A}$$

R_2^T Rule of introducing H:

$$\frac{\vdash A}{\vdash \mathsf{H}A}.$$

To the formula, which is a theorem, these rules allow an addition of one of the above operators, also resulting in a theorem. Logic $\mathbf{K}_t$ is basic, because its axioms and the formulas that result from them are satisfied in each class of interpretation.

Transitivity and linearity of relation R is guaranteed in tense logic **CL**, proposed by N. B. Cocchiarella. This logic is obtained by adding, to the axioms and rules of $\mathbf{K}_t$ the further following axioms:

$$A_5^{TCL}\ \mathsf{FF}A \rightarrow \mathsf{F}A$$

$$A_6^{TCL}\ (\mathsf{F}A \wedge \mathsf{F}B) \rightarrow \mathsf{F}(A \wedge B) \vee \mathsf{F}(A \wedge \mathsf{F}B) \vee \mathsf{F}(\mathsf{F}A \wedge B)$$

$$A_7^{TCL}\ (\mathsf{P}A \wedge \mathsf{P}B) \rightarrow \mathsf{P}(A \wedge B) \vee \mathsf{P}(A \wedge \mathsf{P}B) \vee \mathsf{P}(\mathsf{P}A \wedge B).$$

Axiom A_5^{TCL} guarantees the transitivity of relation R (its equivalent to functor P is provable), whereas the other axioms guarantee so-called right and left linearity of relation R, expressed by the following conditions:

$$\forall_{x,y,z \in W} (xRy \wedge xRz \rightarrow y = z \vee yRz \vee zRy)$$

$$\forall_{x,y,z \in W} (yRx \wedge zRx \rightarrow y = z \vee yRz \vee zRy).$$

Adding another two axioms to the mentioned above:

$$A_8^{TSL}\ \mathsf{G}A \rightarrow \mathsf{F}A$$

$$A_9^{TSL}\ \mathsf{H}A \rightarrow \mathsf{P}A$$

we shall get a so-called logic of the infinite time, described by Dana Scott. It guarantees relation R to meet the following conditions:

$$\forall_{x \in W} \exists_{y \in W} (xRy)$$

$$\forall_{x \in W} \exists_{y \in W} (yRx).$$

Notice, though, that if R is reflexive, then these conditions do not mean the absence of the first and last elements. Only the assumption of the irreflexivity guarantees that with each x there is some new y in a relation, and vice versa.

In turn, by adding to the above axioms, the following axiom:

$$A_{10}^{TP}\ \mathsf{F}A \rightarrow \mathsf{FF}A$$

we get a so-called dense time logic (analogous formula with operator P is derivable), introduced by A. N. Prior. It guarantees the following qualities of relation R:

$$\forall_{x,y\in W}(xRy \to \exists_{z\in W}(xRz \wedge zRy)).$$

However, you can see again that if R is reflexive, then the matter involves a concept of density, different from the one we have previously introduced.

The issue of the branching of the structure defined by R usually is not settled positively, excluding the linearity, but only allowed. For example, Cocchiarella introduces system **CR** that is formed by adding to $\mathbf{K}_t$ axiom A_5^{TCL}. Thus, it allows the absence of linearity from both, right and left sides. On the other hand, Rescher and Urquhart formulated system $\mathbf{K}_b$ that excludes the absence of left linearity, adding to the axioms **CR** axiom A_7^{TCL}.

Within the tense logic, we shall neither define the concept of the proof nor the concept of the theorem. They are analogous to the ones in **CPL**, it is just that the individual formulas can appear in a proof sequence, as special cases of new adequate for the given logic axioms, and while applying new rules of inference.

Let us consider another issue. In the literature, it is emphasised that there is no formula of the tense logic that would define the irreflexivity of relation R (cf. Venema Y. [287], p. 209). On the other hand, if considering a class of such interpretations where relation R is reflexive, then it is possible to demonstrate that the following schemes of formulas are true:

1. $A \to \mathsf{F}A$
2. $A \to \mathsf{P}A$
3. $\mathsf{G}A \to A$
4. $\mathsf{H}A \to A$.

Let us consider e.g. formula 1. Assume that $\mathfrak{I}^{TR}$ is any interpretation where R is reflexive. Take any element $x \in W$ such that $\overline{V}(x,A) = 1$. Since R is reflexive, then there is such $y \in W$ $(y = x)$ that xRy and $\overline{V}(y,A) = 1$. Ergo $\overline{V}(x,\mathsf{F}A) = 1$ and $\overline{V}(x,A \to \mathsf{F}A) = 1$. Then if $\overline{V}(x,A) = 0$, then $\overline{V}(x,A \to \mathsf{F}A) = 1$. Point x is arbitrary, so we get $\mathfrak{I}^{TR} \vDash_x A \to \mathsf{F}A$ for any $x \in W$. It finally results in a conclusion that 1. is a tautology of the class of interpretation with a reflexive relation.

Let us now examine formula 4. Take any element $x \in W$ such that $\overline{V}(x,\mathsf{H}A) = 1$. Then for any $y \in W$ if yRx, then $\overline{V}(y,A) = 1$. Since R is reflexive, then xRx, thus $\overline{V}(x,A) = 1$. Then if $\overline{V}(x,HA) = 0$, then $\overline{V}(x,\mathsf{H}A \to A) = 1$. Since x is arbitrary, we get $\mathfrak{I}^{TR} \vDash_x \mathsf{H}A \to A$ for any $x \in W$. It finally results in a conclusion that 4. is a tautology of the class of interpretation with a reflexive relation.

These are quite disturbing conclusions. The authors of the tense logic wanted to avoid referring of the tense operators to the element in which given formula is interpreted. Thus, from this point of view, the structures with a reflexive relation are not interesting. Therefore, the relation of the temporal precession by means of which we have defined time, without the above consequences cannot be identified with relation R. In order to avoid these problems, the time structure can be considered given only different elements, thus considering the elements that belong to $\leq$, but do not belong to identity relation I.

It would mean the analysis of time substructure $\langle T, < \rangle$. Due to the absence of the conditions imposed on relation R in logic $\mathbf{K}_t$, we can be certain that its axioms, as well as their consequences, will be true in the interpretation with structure $\langle T, < \rangle$. Because of transitivity $\leq$ and $<$, also true will be axiom A_5^{TCL} that guarantees the transitivity of relation R. In turn, in accordance to the properties of the past elements of temporal precession relation $\leq$, which is expressed by condition $i)$ of the definition of time, also $<$ has analogous quality. Therefore, in the interpretations with structures $\langle T, < \rangle$ also A_7^{TCL}, which expresses the left linearity, is true. All the axioms and rules of $\mathbf{K}_t$ and axioms A_5^{TCL}, A_7^{TCL} are, thus, true in the interpretations with time structure $\langle T, \leq \rangle$, as well as in their substructures $\langle T, < \rangle$.

Finally, notice that the remaining axioms expressing e.g. absence of an end A_8^{TSL}, absence of a beginning A_9^{TSL} and density A_{10}^{TP}, if they are true in the interpretation with structure $\langle T, < \rangle$, then they guarantee the absence of a beginning and an end, and the density, corresponding to the concepts of an end, beginning and density, adopted in this study.

Such postulates shall be employed along with the tense logics in the further reconstructions of **RDC**. In the study, while we shall employ the means of tense logics, we will indicate this fact by placing the symbol of a corresponding logic on the right side of the proof step.

Part III Solutions

8 Reconstructions with operator $\mathcal{R}$

8.1 The Reconstruction of F. S. Michael

The intention of the author for the reconstruction of **RDC**, presented in this section, was to be accurate, not based on the premisses reflecting the biases of one of the philosophic schools, and finally presenting the argumentation of Diodorus as conclusive (Michael F. S. [190], p. 229). However, a brief analysis indicates that it does not meet all of these conditions.

For, within the proof layer, this reconstruction omits premiss *D*2. The proof leads to the expected contradiction of the three premisses; however, it is only due to the fact that, on the basis of the classical logic, while expanding a contradictory set of sentences by adding an arbitrary new sentence, we still get a contradictory set of sentences. In consequence, Michael's reconstruction is thus in fact a Diodorus-like reasoning, assuming, following the ancients, that the second premiss had played an active part in the argumentation. This reconstruction is of an informal character, thus it requires certain measures in order to, from a preformal outline, become a strict reconstruction and a formalisation of Diodorus' reasoning. Other issues are some problems arising in relation to Diodorus' definitions of modality themselves.

8.1.1 Preliminaries

Notwithstanding, Michaels's work has also many advantages. Both, the advantages and the disadvantages, will be remarked upon during the discussion on the very reasoning. Its satisfying development requires, as its logical basis, a certain logic enabling performing operations of shifting backwards and forwards in time against arbitrary but invariant point of time called 'now' (denoted as: n), in order to assess the logical values of the sentences. Shifting towards *the future* and *the past* will be presented by Michael by means of arithmetic operations of addition and subtraction of the distances from the constant point in a certain time structure.

This point is fixed; otherwise, if n acted as a nonce word, then the sentence realised at n would have been realised at any point. Whereas in Michael's proposition this is not the case. For the substance of his argumentation is the analysis whether a given sentence is realised earlier than at n.

The above findings substantiate the adopting, as a proof frame, of metric logic $\mathcal{R}_n^+$ with fixed n, which outline we have presented[204]. We shall add two modal operators $\Diamond$, $\Box$ to its language, which should be denoted respectively as *it is necessary that* and *it is possible that*[205]. Of course, these are the functors of syntactic characteristic of the sentence-forming functors of a sentential argument. Furthermore, we shall introduce another symbol p_0 that denotes arbitrary but invariant simple sentence which will become a subject of the reasoning.

The sentential variables in Michael's reconstruction — in compliance with his intention, anyway — will range over set $\mathbf{S_{TD}}$.

The premisses relevant in this reconstruction will take the following form:

$D1: \mathcal{R}_{n-d}(A) \rightarrow \Box A$

if sentence A is realised in the past, then it is necessary,

$D3: \Diamond p_0 \wedge \neg\mathcal{R}_n(p_0) \wedge \neg\mathcal{R}_{n+d}(p_0)$

there is a certain sentence that is possible but neither at n it is realised, nor it will be realised in the future.

The set of the additional premisses employed in the proof consists of, accepted in the ancient times and denoted in the styling corresponding to the chosen language, the following rules:

$S_1: \Box A \rightarrow \mathcal{R}_n(A)$

if it is necessary that A, then A is also realised at n,

$S_2: \exists t\, \mathcal{R}_t(A) \rightarrow \Diamond A$

if sentence A is realised at some point of time, then it is possible that A,

$S_3: \Box\neg A \rightarrow \neg\Diamond A$

if it is necessary that $\neg A$, then it is not possible that A,

---

204 Michael's work suggests that he would have been prone to interpret the realisation operator in a rather stronger explication than the one I have suggested. However, it does not change the basic conclusions that can be drawn from the presented interpretation. Instead, it provides the opportunity for its more accurate formulation.

205 In the described reconstructions both, the necessity and probability functors will be denoted by different symbols in accordance with the symbolism adopted by their authors.

$S_4: \Box A \leftrightarrow \neg \Diamond \neg A$

it is necessary that A if and only if it is not possible that $\neg A$.

8.1.2 Reasoning

In his reconstruction, Michael fortifies premiss $D3$ to premiss $D3'$:

$$\Diamond p_0 \wedge \forall_t \neg \mathcal{R}_t(p_0).$$

The justification for $D3'$ is as follows. From $D3$, it is known that p_0 is neither realised in the present nor will be realised in the future. Let us examine whether it has been realised in the past. By the means of **CPL** and $D3$, we get:

1) $\Diamond p_0$	
2) $\neg \mathcal{R}_n(p_0) \wedge \neg \mathcal{R}_{n+d}(p_0)$	
3) $\mathcal{R}_{n-d}(p_0) \rightarrow \Box p_0$	$D1$
4) $\Box p_0 \rightarrow \mathcal{R}_n(p_0)$	S_1
5) $\mathcal{R}_{n-d}(p_0) \rightarrow \mathcal{R}_n(p_0)$	3), 4), **CPL**
6) $\neg \mathcal{R}_n(p_0) \rightarrow \neg \mathcal{R}_{n-d}(p_0)$	5), **CPL**
7) $\neg \mathcal{R}_n(p_0)$	2), **CPL**
8) $\neg \mathcal{R}_{n-d}(p_0)$	6), 7), **CPL**
9) $\neg \mathcal{R}_{n-d}(p_0) \wedge \neg \mathcal{R}_n(p_0) \wedge \neg \mathcal{R}_{n+d}(p_0)$	8), 2), **CPL**
10) $\forall_d (\neg \mathcal{R}_{n-d}(p_0) \wedge \neg \mathcal{R}_n(p_0) \wedge \neg \mathcal{R}_{n+d}(p_0))$	9), **FOL**
11) $\exists_t \forall_d (\neg \mathcal{R}_{t-d}(p_0) \wedge \neg \mathcal{R}_t(p_0) \wedge \neg \mathcal{R}_{t+d}(p_0))$	10), **FOL**
12) $\forall_t \neg \mathcal{R}_t(p_0)$	A_{R7}, 11), **CPL**
13) $\Diamond p_0 \wedge \forall_t \neg \mathcal{R}_t(p_0)$	1), 12), **CPL**

In the thirteenth step, we obtained $D3'$. This sentence states that sentence p_0 is possible, even though it is not realised at any point of time. Ergo, it implies the existence of possible sentences that will never be true! Having employed such fortified premiss, Michael proceeds to the proper proof which is to demonstrate the incoherence of Diodorus' premisses. Actually, this proof could have begun with step 8).

14) $\mathcal{R}_{n-d}(\neg p_0)$	A_{R2}, 8), **CPL**
15) $\Box \neg p_0$	$D1$, 14), **CPL**
16) $\neg \Diamond p_0$	S_3, 15), **CPL**, which negates 1)

Consequently, the accepted set of the assumptions leads to a contradiction. Since Diodorus accepted these premisses, apart from $D3$ (it played an important role in the proof of step 8)), this very one must be rejected and its negation accepted. This is to bring us right to Diodorus' definition of modality.

8.1.3 Definitions of modality

By negating *D*3 and transforming it by means of the laws of **CPL** and **FOL**, we get the following formula:

$$\Diamond p_0 \rightarrow \mathcal{R}_n(p_0) \vee \exists_d \mathcal{R}_{n+d}(p_0)$$

that corresponds to one side of Diodorus' definition. Michael's reconstruction allows to fortify this implication, for, as we can remember, we have proved by means of stronger premiss $D3'$ (using A_{R7} accepted by Michael) by negating which, in a similar way as the previous one, we get a more general formula:

$$\Diamond p_0 \rightarrow \exists_t \mathcal{R}_t(p_0).$$

Notice that since sentence p_0 has been an invariant but arbitrary sentence which properties we have not employed in the proof, nothing prevents the all foregoing considerations to be generalised by exchanging them for any other sentence, that is, a metavariable.

$$\Diamond A \rightarrow \exists_t \mathcal{R}_t(A).$$

Now, by the virtue of S_2 and the laws of **CPL**, from the above formula the following stronger thesis can be derived:

$$D\Diamond' : \Diamond A \leftrightarrow \exists_t \mathcal{R}_t(A).$$

On the other hand, employing S_4, $A_{\mathcal{R}1}$, $A_{\mathcal{R}2}$, $A_{\mathcal{R}3}$ and the laws of **CPL** and **FOL**, we can surely demonstrate that:

$$\Box A \rightarrow \forall_t \mathcal{R}_t(A)$$

and:

$$\forall_t \mathcal{R}_t(A) \rightarrow \Box A.$$

Composing the two implications, we get a thesis that corresponds to Diodorus' definition of necessity:

$$D\Box' : \Box A \leftrightarrow \forall_t \mathcal{R}_t(A).$$

On the basis of the outlined reconstruction and logic $\mathcal{R}_n^+$ we can, therefore, obtain stronger definitions of the temporal modalities than it was intended by Diodorus.

However, let us get back to the negation of $D3$. Remember, though, that in order to obtain $D3'$ we have employed axiom $A_{\mathcal{R}7}$ that is quite controversial. The negation of $D3$ leads to the following thesis:

$$\Diamond p_0 \rightarrow \mathcal{R}_n(p_0) \vee \exists_d \mathcal{R}_{n+d}(p_0).$$

The converse of this thesis is obtained by means of **FOL** and S_2. While the both implications, generalised into arbitrary sentences, sum up to Diodorus' definition of possibility:

$$D\Diamond: \Diamond A \leftrightarrow \mathcal{R}_n(A) \vee \exists_d \mathcal{R}_{n+d}(A).$$

Furthermore, employing S_4, $A_{\mathcal{R}1}$, $A_{\mathcal{R}2}$, $A_{\mathcal{R}3}$ and the laws of **CPL** and **FOL** we can state that:

$$D\Box: \Box A \leftrightarrow \mathcal{R}_n(A) \wedge \forall_d \mathcal{R}_{n+d}(A).$$

8.1.4 Time structure

In the context of the subject of this study, it is important to settle the issue of which time structure is favoured within the presented theory. Due to the metric logic employed herein, the time structure has been represented by an adequate coordinating structure. However, the employed premisses do not preordain whether this structure is continuous or discrete. Consequently, due to the isomorphism between the coordinating structure and the time structure, also the time structure can be continuous or discrete.

However, for the purposes of the proper defining of the operations of addition and subtraction, the metric logic requires such coordinating structure which guarantees that, while adding or subtracting any number, we shall find a suitable element that constitutes a result of this operation. Thus, this structure does neither have the first element, nor the last one. Therefore, within the time structure that is coordinated, none of the branches has the first, nor the last moment. Ergo, each of its branches constitutes the time with no end and no beginning.

The last important question is the issue whether the presented theory excludes the branching of time. The reconstruction has been presented assuming the use of the tools of thereinbefore outlined logic $\mathcal{R}_n^+$. This means also that we are governed by the presented semantic intuitions.

In the case of reasoning on stronger modal concepts $D\Diamond'$ and $D\Box'$, the time structure cannot be branching. Notwithstanding, it is not only a result of employing axiom $A_{\mathcal{R}7}$ (for we have used it in a very special substitution, which by itself does not require the structure to be linear). The key here is premiss $D1$ that, after the substitution of $\Box$ by a definiens from definition $D\Box'$, is as follows:

$$\mathcal{R}_{n-d}(A) \rightarrow \forall_t \mathcal{R}_t(A).$$

In its case, we can take substitution p and construct such branching model that at only one point in only one branch of it p is true, whereas at all the other points p is false. Since in this model, moment n precedes the branching, then for any d formula $\mathcal{R}_{n-d}(p)$ is true, but it is not true that $\forall_t \mathcal{R}_t(p)$, as in the second of the branches there is at least one moment x at which sentence p is not realised.

Nevertheless, we do not have to circumstantiate such strong concepts as $D\Diamond'$ and $D|Box$. For to Diodorus' concepts correspond notions $D\Diamond$ and $D\Box$. In their proof, we have not employed axiom $A_{\mathcal{R}7}$, and premiss $D1$ takes the following form:

$$\mathcal{R}_{n-d}(A) \rightarrow \mathcal{R}_n(A) \wedge \mathcal{R}_{n+d}(A).$$

In the presence of the established concept of realisation and the introduced semantic intuitions, this premiss does not exclude the branching structure of time. The rest of the premisses, under the both approaches, bear no meaning for the time structure.

However, it is worth noting that a change in the understanding of operator $\mathcal{R}$, a change in the interpretation of position n or a change in the interpretation of the expressions about addition or subtraction could lead to entirely different conclusions.

8.2 Reconstruction of N. Rescher

Rescher states that his reconstruction does not have any historic aspirations. Notwithstanding, in its essence, it aims to be close to the idea of Diodorus, by meeting the following conditions: i) it is to employ the initial ideas of Diodorus, ii) it is to be logically convincing within nowadays implemented standards and iii) it is to achieve the goals assigned to the original reasoning of Diodorus ([243], p. 438). The basic feature distinguishing it from the previously presented reconstruction is the fact that its author employs all of Diodorus' premisses. However, he does it in an interesting manner, interpreting the second premiss in line with the method proposed by Zeller. Rescher's reconstruction, thus, is based on the second interpretation of the premiss number two, that is, $D2'$. In the present subchapter, we shall introduce the original reconstruction, whereas in the next one, the reconstruction that is inspired by this idea.

8.2.1 Preliminaries

In his reconstruction, Rescher does not indicate any determined logical ground. Apart from above-described realisation operator R_t, in the language, though, he employs a two-argument predicate corresponding to the relation of *being earlier* of the moments of time. However, he is not always consistent. In his previous article, where he describes the reconstruction based on the same premisses, for in the key step of the proof he performs an addition operation on the temporal variables as if he considered time in a metric manner [243]. In another part, he outlines his reconstruction with no reference to such operations on the moments, employing, introduced to the language, predicate < ([241], pp. 189–212). This very version of the reconstruction shall be presented herein[206].

The Rescher's reconstruction will be expressed in the language that consists of constants of the classical logic, quantifiers, realisation operators, sentential variables, temporal variables and constant n for arbitrary but determined point corresponding to the present. Additionally, also the predicate of *being earlier* $\leq$ and identity predicate = belong to the language. By the means of the following definitions, we shall also introduce predicates $\neq$, <:

$$DEF^{\neq}: t_1 \neq t_2 \equiv \neg t_1 = t_2$$

$$DEF^{<}: t_1 < t_t \equiv t_1 \neq t_2 \wedge t_1 \leq t_2.$$

In the proof — in transforming the formulas with the predicates or operator $\mathcal{R}_t$ — the use of all the laws of classical logic and the following axiom for the realisation operator is allowed:

$$A_{\mathcal{R}2}: \neg\mathcal{R}_t(A) \leftrightarrow \mathcal{R}_t(\neg A)^{207}.$$

Since Rescher opts for the interpretation of the second premiss of Diodorus in accordance with Zeller, in his proof, there are chronologically arranged modalities. It means that sentences are not simply possible or necessary, but they are possible or necessary at a certain (or at each) point of time: $\Diamond_t A$, $\Box_t A$. This concept of modality is imposed by the interpretation of the second premiss of Diodorus, as stating that the sentence possible at a given point of time is possible at each subsequent point of time.

The last essential remark relates to the range of the sentential variables. According to Rescher, the interpretation of the second premiss of Diodorus

206 The overview of this reconstruction can be found in Jarmużek T. [122].

207 Presumably, Rescher would have been prone to interpret the realisation operator in a similar way he did in his logics of realisation, i.e. $\overline{V}(\mathcal{R}_t(A)) = 1 \Leftrightarrow \overline{V}(t, A) = 1$. It allows to accept the above axiom.

that he has adopted, requires employing sentences determined in time, because for the sentences undetermined it would have been non-intuitive. Here is his example. Sentence 'Tomorrow is Wednesday.' is possible on Tuesday, whereas a day later, i.e. on Wednesday, it is not possible any more ([243], p. 440).

We can suppose that this objection does not relate to the sentences determined in time. For example, sentence '20.05.2013 is May.' is possible on Tuesday, as well as on Wednesday and on any other following weekday. By all means it is a good reason to exclude the temporally indeterminate sentences from the field of considerations, but only when the concept is *a priori* assigned Diodorian features, i.e. that the possible sentence is or will be true. However, if we do not proceed in this manner — and, actually, we should not, since Diodorus' definitions are to be the very outcome of the reasoning — then there is no reason not to state that the use of the sentences determined in time is also non-intuitive.

Let us consider e.g. sentence 'In 1410, the Knights of the Cross will win the Battle of Grunwald.' There are no contraindications to state that this sentence was possible in 1409. However, two years later, in the temporal sense, it ceased to be possible, for all the states of affairs from the year 1410 have passed into the history and there had been no verifier that would make them true. This issue, though, is not so clear. However, it is not a question of the choice of sentences — as indicated above — but of the premiss itself.

Now we shall present the specific axioms from Rescher's theory. Let us begin with the premisses of Diodorus:

$D1^R : (\mathcal{R}_{t_1}(A) \wedge t_1 < t_2) \rightarrow \Box_{t_2} A$

if sentence A is realised at time t_1, than at any later time it is necessary.

Rescher makes use of the above stronger version of Diodorus' premiss which states that the sentences become necessary at any subsequent moments, with no reference to the present. For the first premiss of Diodorus, in his language would have taken the following form:

$D1 : (\mathcal{R}_{t_1}(A) \wedge t_1 < n) \rightarrow \Box_n A$

if sentence A is realised in the past, then it is necessary in the present.

The further premisses of Diodorus take the following form:

$D2' : (\Diamond_{t_1} A \wedge t_1 < t_2) \rightarrow \Diamond_{t_2} A$

if sentence A is possible at time t_1, then it is possible at later time,

$D3 : \Diamond_n p_0 \wedge (n \leq t_1 \rightarrow \neg \mathcal{R}_{t_1}(p_0))$

there is such sentence p_0 that is possible now, but at any other time it is not realised.

In addition to these axioms, Rescher also assumes that time structure $\langle T, \leq \rangle$ is quasi-linear (no antisymmetry condition — however, it is present in the very definition of time); it has been reflected in the following axioms ([241], p. 191).

$$L1:\ t_1 \leq t_1$$
$$L2:\ t_1 \leq t_2 \wedge t_2 \leq t_3 \rightarrow t_1 \leq t_3$$
$$L3:\ t_1 \leq t_2 \vee t_1 = t_2 \vee t_2 \leq t_1$$

Additionally imposing the condition of the absence of the last elements[208].

$$L4:\ \exists_{t_2}\, t_1 < t_2$$

Notice that $L2$, due to the properties of strict order relation <, results in $L2'$:

$$L2':\ t_1 < t_2 \wedge t_2 < t_3 \rightarrow t_1 < t_3.$$

The remaining required principles concern the relations between the modalities. Aristotelian definitions of possibility and necessity are essential:

$$S_1:\ \Diamond_{t_1} A \leftrightarrow \neg \Box_{t_1} \neg A$$
$$S_2:\ \Box_{t_1} A \leftrightarrow \neg \Diamond_{t_1} \neg A.$$

However, if we aim to circumstantiate Diodorus' definitions, another rule is required; it states that the sentences realised in the future relative to n are possible:

$$S_3:\ \exists_{t_1}(n \leq t_1 \wedge \mathcal{R}_{t_1}(A)) \rightarrow \Diamond_n A.$$

This rule can take a more general form, acknowledging the above relation not only in regard to point n, but also to any moments:

$$S_3':\ \exists_{t_1}(t_1 \leq t_2 \wedge \mathcal{R}_{t_1}(A)) \rightarrow \Diamond_{t_2} A.$$

8.2.2 Reasoning

Having acquired the above formal tools, we can now proceed to the formal presentation of the reconstruction of **RDC** propounded by Nicholas Rescher.

---

208 Within the original proof of Rescher, it is assumed that the time structure is a strict linear order ([241], pp. 193–195) and that for any t_1 there are such two different t_2, t_3 that $t_1 < t_2$ and $t_1 < t_3$, which is equivalent to $L4$, with all other assumptions. However, in order to simplify the proof, I assume a partial linear order of time, which will enable the direct use of the reflexivity of the relation of temporal precession in $D3$.

1) $(\mathcal{R}_{t_1}(\neg A) \wedge t_1 < t_2) \rightarrow \Box_{t_2}\neg A$ — $D1^R$
2) $\neg\Box_{t_2}\neg A \rightarrow \neg(\mathcal{R}_{t_1}(\neg A) \wedge t_1 < t_2)$ — 1), **CPL**
3) $\Diamond_{t_2}A \leftrightarrow \neg\Box_{t_2}\neg A$ — S_1
4) $\Diamond_{t_2}A \rightarrow \neg\Box_{t_2}\neg A$ — 3), **CPL**
5) $\Diamond_{t_2}A \rightarrow \neg(\mathcal{R}_{t_1}\neg A \wedge t_1 < t_2)$ — 2), 4), **CPL**
6) $\neg(\mathcal{R}_{t_1}(\neg A) \wedge t_1 < t_2) \rightarrow (t_1 < t_2 \rightarrow \neg\mathcal{R}_{t_1}(\neg A))$ — **CPL**
7) $\Diamond_{t_2}A \rightarrow (t_1 < t_2 \rightarrow \neg\mathcal{R}_{t_1}(\neg A))$ — 5), 6), **CPL**
8) $(\Diamond_{t_2}A \wedge t_1 < t_2) \rightarrow \neg\mathcal{R}_{t_1}(\neg A)$ — 7), **CPL**
9) $\neg\mathcal{R}_{t_1}(\neg A) \rightarrow \mathcal{R}_{t_1}(A)$ — $A_{\mathcal{R}2}$, **CPL**
10) $(\Diamond_{t_2}A \wedge t_1 < t_2) \rightarrow \mathcal{R}_{t_1}(A)$ — 8), 9), **CPL**
11) $\Diamond_n p_0$ — $D3$, **CPL**
12) $\exists_{t_1} n < t_1$ — $L4$, **FOL**
13) $n < a$ — 12), **FOL**
14) $(\Diamond_n p_0 \wedge n < t_2) \rightarrow \Diamond_{t_2} p_0$ — $D2'$, **FOL**
15) $(\Diamond_n p_0 \wedge n < a) \rightarrow \Diamond_a p_0$ — 14), **FOL**
16) $\Diamond_n p_0 \wedge n < a$ — 11, 13), **CPL**
17) $\Diamond_a p_0$ — 15), 16), **CPL**
18) $(\Diamond_a p_0 \wedge t_1 < a) \rightarrow \mathcal{R}_{t_1}(p_0)$ — 10), **FOL**
19) $(\Diamond_a p_0 \wedge n < a) \rightarrow \mathcal{R}_n(p_0)$ — 18), **FOL**
20) $\Diamond_a p_0 \wedge n < a$ — 17), 13), **CPL**
21) $\mathcal{R}_n(p_0)$ — 19), 20), **CPL**
22) $n \leq n$ — $L1$, **FOL**
23) $n \leq t_1 \rightarrow \mathcal{R}_{t_1}(\neg p_0)$ — $D3$, **CPL**
24) $n \leq n \rightarrow \mathcal{R}_n(\neg p_0)$ — 23), **FOL**
25) $\mathcal{R}_n(\neg p_0)$ — 24), 22), **CPL**
26) $\mathcal{R}_n(\neg p_0) \rightarrow \neg\mathcal{R}_n(p_0)$ — $A_{\mathcal{R}2}$, **FOL**
27) $\neg\mathcal{R}_n(p_0)$ — 26), 25), **CPL**, contradiction with step 21)

The above proof could have been carried out in a quite different manner, in accordance to the strategy suggested by Jaakko Hintikka in his outline of the reconstruction of the reasoning of Diodorus (cf. [104], pp. 179–213, Gaskin R. [80], pp. 236–241). However, since Hintikka does not introduce any formal proof there, in order to present the line of reasoning close to the one he proposed, we shall employ Rescher's premisses.

1) $\Diamond_n p_0$ — $D3$, **CPL**
2) $\exists_{t_2} n < t_2$ — $L4$, **FOL**
3) $n < a$ — 2), **FOL**
4) $n \leq a$ — 3), $DEF^<$, **FOL**

5)	$\forall_{t_1}(n \leq t_1 \rightarrow \neg\mathcal{R}_{t_1}(p_0))$	$D3$, **CPL**
6)	$n \leq a \rightarrow \neg\mathcal{R}_a(p_0)$	5), **FOL**
7)	$\neg\mathcal{R}_a(p_0)$	6), 4), **CPL**
8)	$\neg\mathcal{R}_a(p_0) \leftrightarrow \mathcal{R}_a(\neg p_0)$	$A_{\mathcal{R}2}$, **FOL**
9)	$\neg\mathcal{R}_a(p_0) \rightarrow \mathcal{R}_a(\neg p_0)$	8), **CPL**
10)	$\mathcal{R}_a(\neg p_0)$	9), 7), **CPL**
11)	$\exists_{t_2}\, a < t_1$	$L4$, **FOL**
12)	$a < b$	11), **FOL**
13)	$\mathcal{R}_a(\neg p_0) \wedge a < b \rightarrow \Box_b \neg p_0$	$D1^R$, **FOL**
14)	$\mathcal{R}_a(\neg p_0) \wedge a < b$	10), 12), **CPL**
15)	$\Box_b \neg p_0$	13), 14), **CPL**
16)	$\Diamond_b p_0 \leftrightarrow \neg \Box_b \neg p_0$	S_1, **FOL**
17)	$\neg \Diamond_b p_0 \leftrightarrow \Box_b \neg p_0$	16), **CPL**
18)	$\Box_b \neg p_0 \rightarrow \neg \Diamond_b p_0$	17), **CPL**
19)	$\neg \Diamond_b p_0$	18), 15), **CPL**
20)	$n \leq a \wedge a \leq b \rightarrow n \leq b$	$L2$, **FOL**
21)	$n < a \wedge a < b \rightarrow n < b$	20), $L2'$
22)	$n < a \wedge a < b$	3), 12), **CPL**
23)	$n < b$	21), 22), **CPL**
24)	$\Diamond_n p_0 \wedge n < b \rightarrow \Diamond_b p_0$	$D2'$, **FOL**
25)	$\Diamond_n p_0 \rightarrow \Diamond_b p_0$	23), 24), **CPL**
26)	$\neg \Diamond_b p_0 \rightarrow \neg \Diamond_n p_0$	25), **CPL**
27)	$\neg \Diamond_n p_0$	26), 19), **CPL**, contradiction with step 1)

Similarly to the previous reconstruction, also in this one, premiss $D3$ has played a substantial role. Thus, considering it as an indirect assumption and employing the relevant laws of the classical logic, we can assume its negation.

8.2.3 Definitions of modality

Since in the proof nothing has been assumed about sentence p_0 apart from what $D3$ has established, the whole reasoning can be generalised into arbitrary formula by inserting a metavariable. In that case, let us negate $D3$ in a more general form:

$$\neg(\Diamond_n A \wedge \forall_{t_1}(n \leq t_1 \rightarrow \neg\mathcal{R}_{t_1}(A))).$$

By transforming this formula based on the relevant laws of **CPL** and **FOL**, we get the following formula:

$$\Diamond_n A \rightarrow \exists_{t_1}(n \leq t_1 \wedge \mathcal{R}_{t_1}(A)).$$

Composing it with S_3, we obtain Diodorus' definition:

$$D\Diamond\colon\ \Diamond_n(A) \leftrightarrow \exists_{t_1}(n \leq t_1 \wedge \mathcal{R}_{t_1}(A))$$

which states that a sentence is possible in the present if and only if it is realised in the present or at any other future moment. Employing the relevant laws of **CPL**, **FOL**, principle S_2 and axiom $A_{\mathcal{R}2}$, we can prove the formula that expresses Diodorus' definition of necessity:

$$D\Box\colon\ \Box_n A \leftrightarrow \forall_{t_1}(n \leq t_1 \rightarrow \mathcal{R}_{t_1}(A)).$$

This definition states that at the present moment a sentence is necessary if and only if it is now, and always in the future, realised.

In terms of their content, the obtained definitions correspond to the remains from the data about the Diodorus' concept. Naturally, the above reasoning can be generalised. For point of time n was a determined but arbitrary point and in the proof nothing has been assumed about it, whereas the time structure did not contain the last element. This means that in $D3$ instead of n there could have been any other moment of time placed. This, of course, leads to the more general definitions of modality:

$$D\Diamond'\colon\ \Diamond_{t_1} A \leftrightarrow \exists_{t_2}(t_1 \leq t_2 \wedge \mathcal{R}_{t_2}(A))$$
$$D\Box'\colon\ \Box_{t_1} A \leftrightarrow \forall_{t_2}(t_1 \leq t_2 \rightarrow \mathcal{R}_{t_2}(A)).$$

8.2.4 Time structure

On the surface, the issue of the time structure seems quite clear — the author's intention was for it to be a linear structure. In the above presentation, we have assumed the linearly ordered structure, whereas, in line with the idea of Rescher, we could have assumed a strictly linearly ordered structure. Apparently, in the proof there is no totality axiom ($L3$), thus it seems that the linearity is not necessary. Therefore, let us examine the proof and its consequences.

The proof evidently requires the time structure not to have a last element (in the both versions of the proof we have employed axiom $L4$) — therefore, the time with no end is needed. The issue of the first element, as well as of the discreteness, density or continuity of the time are herein irrelevant.

Let us consider the premisses. In order to reduce the modalities present in $D1^R$ and in $D2'$, we need to employ definitions $D\Box'$ and $D\Diamond'$ — due to the fact that they consist of the temporal variables instead of constant n. Let us reduce the functor of necessity in premiss $D1^R$. By means of $D\Box'$, we get:

$$(\dagger)\ \mathcal{R}_{t_1}(A) \wedge t_1 < t_2 \rightarrow \forall_{t_3}(t_2 \leq t_3 \rightarrow \mathcal{R}_{t_3}(A)).$$

This formula states, among other things, that if a sentence at a certain moment is true (with Rescherian understanding of the functor of realisation), then it is true at all the subsequent moments. It does not exclude branching, but the sentences will carry the same values within the presumptive branches. If we carry out a similar operation of reduction on premiss $D2'$, then the resulting formula seems to exclude branching, since along with the passage of time we lose an access to some moments. However, the prior formula guarantees that, within the parallel branches, these sentences will carry the same values.

In the theory of Rescher, we come across an interesting situation. Well, the theory itself does not exclude the branching; however, these branches add nothing new, as the logical value of the sentences is determined at the preceding moments. Taking any sentence A and any moment from time structure t_1, we know — according to Rescher's interpretation — that sentence A is true or false at t_1. Ergo, it is true that $\mathcal{R}_{t_1}(A)$, or it is true that $\mathcal{R}_{t_1}(\neg A)$. By the virtue of $(\dagger)$ for each t_2 if $t_1 < t_2$, then it is true that $\mathcal{R}_{t_2}(A)$, or for each t_2 if $t_1 < t_2$, then it is true that $\mathcal{R}_{t_2}(\neg A)$, regardless of the quantity of the time branches in the future.

8.3 Calculation of moments

Now we shall present a reconstruction that employs the idea of Zeller and Rescher relating to the second premise of Diodorus, and that, in terms of the very strategy of the proof, is very similar to the previous reconstruction. However, it differs in terms of the used language and some of the consequences. For within this reconstruction, time is not considered a linear structure. Furthermore, the time is regarded metrically.

8.3.1 Preliminaries

In the presented reconstruction, we shall employ the axioms and their interpretations from subchapter 7.4. Let us remind that this interpretation is weaker than the one proffered by Rescher. Moreover, the issue of temporal precession of the moments, within this approach, is represented by means of addition or subtraction of the distances within the coordinating structure. The fact that two moments within a structure are in relation $t_1 < t_2$ means that there is such number $d \neq 0$, called the distance, that its addition to the number representing point t_1 within the coordinating structure results in a number representing subsequent point t_2: $f(t_1) + d = f(t_2)$. Note that the sentential variables will be ranging over the sentences determined in time.

Diodorus' premisses, according to Rescher's interpretation, expressed in the new language, take the following form:

$$D1: \mathcal{R}_{t-d}(A) \rightarrow \Box_t A$$

if sentence A is realised before temporal point t, then at time t it is necessary,

$$D2': \Diamond_t A \rightarrow \Diamond_{t+d} A$$

if sentence A is possible at temporal point t, then it is possible at subsequent point,

$$D3: \Diamond_n p_0 \wedge \neg \mathcal{R}_n p_0 \wedge \neg \exists_d \mathcal{R}_{n+d}\, p_0$$

there is such sentence p_0 that is possible now, but neither is it realised now, nor will it be realised at any future point of time.

Premiss $D1$, similarly as in Rescher's reconstruction, is formulated in a more general manner than its ancient version. The initial premiss stated that the sentences realised in the future are necessary in the present. Notice that according to the new interpretation of operator $\mathcal{R}_i$, being realised at a given point does not mean being *true at this point*. For this very reason, we do not assume that distance d must be different from zero, especially since for each moment within the coordinated structures there exist moments, both, earlier and later ones.

In addition to Diodorus' premisses, in the proof, there will also be required another Aristotle's principles defining possibility by necessity, by means of negation, and a reverse rule:

$$S_1: \Diamond_t A \leftrightarrow \neg \Box_t \neg A$$

$$S_2: \Box_t A \leftrightarrow \neg \Diamond_t \neg A.$$

Moreover, for the proof of Diodorus' definition, there is a following rule needed:

$$S_3: \mathcal{R}_n(A) \vee \exists_d \mathcal{R}_{n+d}(A) \rightarrow \Diamond_n A$$

and axioms $A_{\mathcal{R}2}$, $A_{\mathcal{R}6}$.

8.3.2 Reasoning

Having stocked up on the above formal tools, we can now proceed to the reasoning reconstruction itself.

1) $\mathcal{R}_{t-d}(\neg p_0) \rightarrow \Box_t \neg p_0$ — $D1$
2) $\neg \Box_t \neg p_0 \rightarrow \neg \mathcal{R}_{t-d}(\neg p_0)$ — 1), **CPL**
3) $\Diamond_t p_0 \rightarrow \neg \Box_t \neg p_0$ — S_1, **CPL**
4) $\Diamond_t p_0 \rightarrow \neg \mathcal{R}_{t-d}(\neg p_0)$ — 3), 2) and **CPL**

5) $\Diamond_t p_0 \to \mathcal{R}_{t-d}(p_0)$	4), $A_{\mathcal{R}2}$ and **CPL**
6) $\Diamond_n p_0$	*D*3, **CPL**
7) $\neg\mathcal{R}_n(p_0) \wedge \neg\exists_d \mathcal{R}_{n+d}(p_0)$	*D*3, **CPL**
8) $\Diamond_n p_0 \to \Diamond_{n+d} p_0$	*D*2, **FOL**
9) $\Diamond_{n+d} p_0$	8), 6) and **CPL**
10) $\Diamond_{n+d} p_0 \to \mathcal{R}_{(n+d)-d}(p_0)$	5), **FOL**
11) $\mathcal{R}_{(n+d)-d}(p_0)$	10), 9) and **CPL**
12) $\mathcal{R}_n(p_0)$	$A_{\mathcal{R}6}$, 11) and **CPL**
13) $\neg\mathcal{R}_n(p_0)$	7) and **CPL**, which results in contradiction with step 12)

8.3.3 Definitions of modality

In accordance with Diodorus' strategy, the contradiction is to cause the negation of premiss *D*3 that, based on the laws of **CPL**, results in one side of Diodorus' definition:

$$\Diamond_n p_0 \to \mathcal{R}_n(p_0) \vee \exists_d \mathcal{R}_{n+d}(p_0).$$

Composing it with S_3 we get an equivalence fully corresponding to Diodorus' definition:

$$D\Diamond: \Diamond_n p_0 \leftrightarrow \mathcal{R}_n(p_0) \vee \exists_d \mathcal{R}_{n+d}(p_0)$$

sentence p_0 is possible in the present if and only if it is realised in the present, or it is realised in some future moment.

Similarly as in the previous reconstructions, in the proof there has nothing been assumed either about sentence p_0 or about n. The obtained definition can be thus generalised:

$$D\Diamond': \Diamond_t A \leftrightarrow \mathcal{R}_t(A) \vee \exists_d \mathcal{R}_{t+d}(A).$$

By means of the laws of **CPL**, **FOL**, S_2 and $A_{\mathcal{R}2}$ the formula defining the operator of necessity, that also expresses the intentions analogous to Diodorus' conception, can be obtained:

$$D\Box': \Box_t A \leftrightarrow \mathcal{R}_t(A) \wedge \forall_d \mathcal{R}_{t+d}(A).$$

Thus, both the modal operators can be defined in accordance with Diodorus' concepts in language $\mathcal{R}_n^+$ in a more general manner.

8.3.4 Time structure

In the above reconstruction of the proof and the definition of Diodorus, we have employed the language that requires all the branches of time to have no end and no beginning, and assumes an existence of the coordinating structure. The issues of discreteness or continuity of the time are in this case undetermined. The logical tools themselves also do not exclude the branching into the future.

A separate question is whether, with the modality interpretation that Diodorus' definitions lead to, the remaining premisses allow the branching of time. Let us try to interpret the premisses accepted by Diodorus, by means of $D\Diamond'$ and $D\Box'$. They take the following form:

$$D1\ \mathcal{R}_{t-d_1}(A) \rightarrow \mathcal{R}_t(A) \wedge \forall_{d_2} \mathcal{R}_{t+d_2}(A)$$

$$D2'\ \mathcal{R}_t(A) \vee \exists_{d_1} \mathcal{R}_{t+d_1}(A) \rightarrow \mathcal{R}_{t+d_2}(A) \vee \exists_{d_3} \mathcal{R}_{(t+d_2)+d_3}(A).$$

For formula $D1$ in each branching time structure $\mathcal{R}_n^+$ there can be indicated an interpretation where it is false, as the moment referred to in formula $\mathcal{R}_{t-d_1}(A)$ may cease to be accessible after passing the branching. Even though the same logical tools allow the branching into the future, the additional premisses exclude such possibility.

Notice that in the analysis we have given a miss to a certain quality of logics $\mathcal{R}_n^+$ — from the constant interpretation of point n. If, however, we adhere to it, then it would have not been possible to reduce premisses $D1$ and $D2'$. We could consider the versions of these premisses substituted with n, repeat the reconstruction of the reasoning by rewriting the given reasoning substituted with constant n and finally make a reduction by applying definitions $D\Diamond$ and $D\Box$. The result would not, however, be analogous, for point n would have been located before the branching. In this limited approach, the time could have been branching, due to the remarks on moment n from subchapter 7.4.

9 Other reconstructions

9.1 Reconstruction of A. N. Prior

The idea on which the reconstruction of Arthur R. Prior is based has been presented in his various works (e.g. [228]) and, in the end, in his book *Past, Present and Future* [229]. The logical tools employed in Prior's reconstruction are tense logic $\mathbf{K}_t$ and certain additional assumptions. So the reasoning employs the sentences undetermined in time. We shall present this reconstruction in a slightly modified form, at the same time sustaining the whole idea of its author[209].

9.1.1 Preliminaries

To the language of logic $\mathbf{K}_t$ there should be added two new one-argument modal operators $\Box$ and $\Diamond$ that create formulas from formulas. Besides that, we employ additional premisses which will be discussed consecutively. In accordance with Prior's interpretation, Diodorus' premisses, expressed in the described language, take the following form:

$$D1\colon\ \mathsf{P}A \rightarrow \Box\mathsf{P}A$$

if it has been the case that A, then it has been necessary that A,

$$D2\colon\ \Box(A \rightarrow B) \rightarrow (\Diamond A \rightarrow \Diamond B)$$

if it has been necessary that sentence A implicates sentence B and that sentence A is possible, then sentence B is also possible; this premiss is to correspond to the first premiss of Diodorus, i.e. that the impossible sentence cannot result from a possible one, that is, after the appropriate transitions, to formula: $(\neg \Diamond B \wedge \Diamond A) \rightarrow \neg \Box (A \rightarrow B)$[210]

$$D3\colon\ \Diamond p_0 \wedge \neg p_0 \wedge \neg \mathsf{F} p_0$$

there is such sentence p_0 that is possible, but neither is nor will be true.

209 The complex metalogic research on the logic and the time structures for the logic employed in Prior's proof have been carried out in the work of Jarmużek T. and Pietruszczak A. [120]. In this book, we present only certain aspects, mainly from the philosophical perspective.

210 Instead of such formulation of the second premiss of Diodorus and further employed principle of generalisation for necessity, we can apply the rule of monotonicity for the possibility. See [74, 120].

Premiss *D*1, similarly as in Rescher's reconstruction, is formulated in a more general manner than its ancient version. The initial premiss stated that the sentences realised in the future are necessary in the present.

Apart from Diodorus' premisses, in the proof there will also be another Aristotle's principle required; it defines the necessity by means of the possibility and the negation:

$$S_1\colon \Box A \leftrightarrow \neg \Diamond \neg A$$

and some additional premisses that we will have employed in the proof:

$$P_1\colon A \wedge \mathsf{G}A \to \mathsf{PG}A$$

$$P_2\colon \text{Rule of introducing } \Box\colon$$

$$\frac{\vdash A}{\vdash \Box A}.$$

This rule should be understood in such way that if formula A is a theorem of $\mathbf{K}_t$, then also formula $\Box A$ is a theorem.

Let us also add that the theorem of logic $\mathbf{K}_t$ is the following formula:

$$Th_{\mathbf{K}_t}\colon \mathsf{HF}A \to \neg\mathsf{PG}\neg A.$$

Additionally, for the proof of Diodorus' definition, the following rule is required:

$$S_2\colon A \vee \mathsf{F}A \to \Diamond A.$$

9.1.2 Reasoning

Having stocked up on the above formal tools, now we can proceed to the reconstruction of the reasoning.

1) $\Diamond p_0$	$D3$, **CPL**
2) $\Box(p_0 \to \mathsf{HF}p_0)$	A_3^T, P_2
3) $\Box(p_0 \to \mathsf{HF}p_0) \to (\Diamond p_0 \to \Diamond\mathsf{HF}p_0)$	$D2$
4) $\Diamond p_0 \to \Diamond\mathsf{HF}p_0$	3, 2) and **CPL**
5) $\Diamond\mathsf{HF}p_0$	4), 1) and **CPL**
6) $\neg p_0 \wedge \neg\mathsf{F}p_0$	$D3$, **CPL**
7) $\neg p_0 \wedge \mathsf{G}\neg p_0$	6), DEF^{F} and **CPL**
8) $\neg p_0 \wedge \mathsf{G}\neg p_0 \to \mathsf{PG}\neg p_0$	P_1
9) $\mathsf{PG}\neg p_0$	8), 7) and **CPL**
10) $\Box\mathsf{PG}\neg p_0$	$D1$, 9) and **CPL**
11) $\Box\mathsf{PG}\neg p_0 \to \neg\Diamond\neg\mathsf{PG}\neg p_0$	S_1, **CPL**

12) $\neg\Diamond\neg\mathsf{PG}\neg p_0$ — 11), 10), **CPL**
13) $\Box(\mathsf{HF}p_0 \to \neg\mathsf{PG}\neg p_0)$ — $Tw_{\mathbf{K}_t}$, P_2
14) $\Box(\mathsf{HF}p_0 \to \neg\mathsf{PG}\neg p_0) \to (\Diamond\mathsf{HF}p_0 \to \Diamond\neg\mathsf{PG}\neg p_0)$ — $D2$
15) $\Diamond\mathsf{HF}p_0 \to \Diamond\neg\mathsf{PG}\neg p_0$ — 14), 13) and **CPL**
16) $\neg\Diamond\neg\mathsf{PG}\neg p_0 \to \neg\Diamond\mathsf{HF}p_0$ — 15) and **CPL**
17) $\neg\Diamond\mathsf{HF}p_0$ — 16, 12) and **CPL**, contradiction with step 5).

9.1.3 Definitions of modality

Similarly as in the previous reconstructions, there has nothing been assumed about variable p_0. Ergo, the whole reasoning can be generalised into any formula A. Rejecting premiss $D3$ and accepting its negation, after the application of the laws of **CPL**, we get one side of Diodorus' definition:

$$\Diamond A \to A \vee \mathsf{F}A.$$

Composing it with S_2, we can now derive the full content of $D\Diamond$:

$$D\Diamond\colon\ \Diamond A \leftrightarrow A \vee \mathsf{F}A$$

sentence A is possible if and only if it is or will be true.

Employing the laws of **CPL**, $\mathbf{K}_t$ and S_1 allows to derive from $D\Diamond$ the definition of necessity:

$$D\Box\colon\ \Box A \leftrightarrow A \wedge \mathsf{G}A$$

which states that a sentence is necessary if and only if it is and always will be true.

9.1.4 Time structure

Notice that, apart from the additional premisses, the whole reasoning has been reconstructed within logic $\mathbf{K}_t$. Ergo, assuming that the relations of accessibility equate to the relation within time substructure $\langle T, < \rangle$, it does not impose any new qualities on the time structure.

Let us now examine the additional premisses. Rule of addition of necessities P_2 that we have employed, along with the obtained temporal characteristics of the necessity, states that if formula A is a theorem, then formula $A \wedge \mathsf{G}A$ is also a theorem. Since in logic $\mathbf{K}_t$ there is a rule of adding operator G to the theorems, the rule employed in the reasoning can be derived from it and from **CPL**.

On the other hand, the positive, that is, not rejected Diodorus' premisses $D1$, $D2$, after substituting the modal operators present in them with their *definiens*, will take the following form:

$$D1': \mathsf{P}A \rightarrow \mathsf{P}A \wedge \mathsf{GP}A$$
$$D2': (A \rightarrow B) \wedge \mathsf{G}(A \rightarrow B) \rightarrow (A \vee \mathsf{F}A \rightarrow B \vee \mathsf{F}B).$$

The second of them is provable by the virtue of **CPL** and e.g. the law of $\mathbf{K}_t$ which states that $\mathsf{G}(A \rightarrow B) \rightarrow (\mathsf{F}A \rightarrow \mathsf{F}B)$. It means that it also does not require any additional assumptions about the relation of accessibility.
In turn, problem $D1'$ actually comes down to investigating whether formula

$$D1'': \mathsf{P}A \rightarrow \mathsf{GP}A$$

imposes any additional requirements for the time structure, for it is possible, by means of the laws of **CPL**, to prove straightforwardly $D1'$ and vice versa.

Let us examine this problem from the semantic side. Take any interpretation $\mathfrak{I}^T = \langle T, <, V\rangle$ and assume that $D1''$ at certain $t_1 \in T$ is not fulfilled: $\mathfrak{I}^T \not\models_{t_1} \mathsf{P}A \rightarrow \mathsf{GP}A$. Employing the presented above definitions, we get $\overline{V}(t_1, \mathsf{P}A \rightarrow \mathsf{GP}A) = 0$. Accordingly, $\overline{V}(t_1, \mathsf{P}A) = 1$ and $\overline{V}(t_1, \mathsf{GP}A) = 0$.

From the first part of the conjunction, we get that there is such $t_2 \in T$ that $t_2 < t_1$ and $\overline{V}(t_2, A) = 1$. Whereas from the second part of the conjunction, we get that there exists such $t_3 \in T$ that $t_1 < t_3$ and $\overline{V}(t_3, \mathsf{P}A) = 0$. If so, then for any $t_4 \in T$ if $t_4 < t_3$, then $\overline{V}(t_4, A) = 0$.

However, it is not true, for relation $<$ is transitive, consequently $t_2 < t_3$, ergo, $\overline{V}(t_2, A) = 0$, what would contradict with the assumption.

It follows that for the truthness of $D1'$ the transitivity of the relation, which is guaranteed in the time structure, is sufficient.

On the other hand, take now any structure $\langle T, R\rangle$ and assume that relation R is not transitive. Ergo, there exist such moments t_1, t_2, t_3 that $t_1 R\, t_2$ and $t_2 R\, t_3$, but it is not the case that $t_1 R\, t_3$. Now, let us take letter p and define such valuation V that $V(t_1, p) = 1$, whereas for the remaining $t \in T$, $V(t, p) = 0$.

By the definition of valuation, it follows that $V(t_2, \mathsf{P}p) = 1$ and $V(t_3, \mathsf{P}p) = 0$, ergo, $V(t_2, \mathsf{GP}p) = 0$. And hence $V(t_2, \mathsf{P}p \rightarrow \mathsf{GP}p) = 0$, thus we have such model with non–transitive relation $\mathfrak{I}^T = \langle T, R, V\rangle$ and such formula A that $\mathfrak{I}^T \not\models_t \mathsf{P}A \rightarrow \mathsf{GP}A$ for some $t \in T$.

Therefore, if assuming the truthness of formula $D1''$, then a transitivity of relation R is a necessary condition.

Thus, accepting premiss $D1$ at Diodorus' definitions requires only the time structures to be transitive, what in the case of $\langle T, <\rangle$ is met by the definition.

The last premiss of the presented reasoning that may require additional assumptions about the time structure is P_1:

$$A \wedge \mathsf{G}A \rightarrow \mathsf{PG}A.$$

This premiss seems non-controversial. However, it actually carries quite complex information[211]. It is worth to notice that its brief analysis seems to exclude the branching of time into the future. We shall present it on graphical models.

Black points will denote moments of time; we assume that the lines joining the points are oriented correspondingly to the reading direction. It means that — unless marked otherwise — the point on the left side stays in relation with the point on the right side that is connected to it with the line.

On the first figure, we present only a part of the model of time in which the branching occurs. However, in the model based on such structure, premiss P_1 does not have to be true.

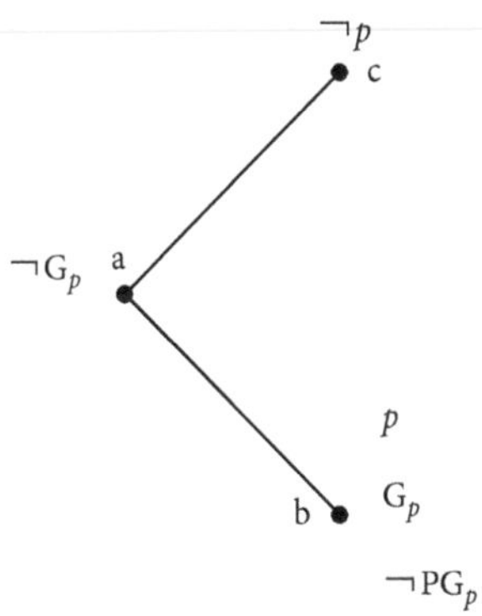

At point a — it does not have to be the first point, as mentioned above we consider here a fragment of a structure — the branching occurs. Now assume that letter p is true at point b and that it will be true at each point from the future b. Then at b it is true that $p \wedge \mathsf{G}p$. Assume, however, that sentence p is not true at point c. Ergo, at point a it is not true that $\mathsf{G}p$. Hence at b it is true that $\neg\mathsf{PG}p$. At point b formula $p \wedge \mathsf{G}p \rightarrow \mathsf{PGP}$ is, thus, not true.

Consequentially, for some sentence A, within each time structure containing such substructure as is shown on the figure, premise P_1 is false. Therefore, if in the proof we assume P_1, then we must exclude the branches of time that are similar

211 A detailed reflection about it can be found in the works of [120, 284]

to those on the figure. Does it mean, though, that the time structure has to be linear?

The answer is: no. In the work [120] such branching structure has been presented in which premise P_1 is true.

Take a set of negative integers and zero $A = \{0, -1, -2, \ldots\}$. Then let us define another two sets $B_1 = \{x : x = \frac{1}{n}{}_1, n \in \mathbb{N} \setminus \{0\}\}$ and $B_2 = \{x : x = \frac{1}{n}{}_2, n \in \mathbb{N} \setminus \{0\}\}$. This set contains the fractions with numerators 1 and which denominators are natural numbers. The only difference between sets B_1 and B_2 is that their elements have different lower indices: $\frac{1}{n}{}_1$ or $\frac{1}{n}{}_2$. Let us now consider regular relation of *being smaller* defined on a set of rational numbers (denoted simply as R). Relation R is, among other things, irreflexive, i.e. $\forall_x \neg xRx$. If on its basis we define relation of *immediate smaller predecessor* R' (for short: *immediate predecessor*) by formula: $xR'y \Leftrightarrow xRy \& \neg\exists_z(xRz \& zRy)$, then all the defined sets have the following properties:

1) each element has its immediate predecessor
2) the number zero is not an immediate predecessor of any element.

On the basis of the above framework, we can draw the following branching time structure.

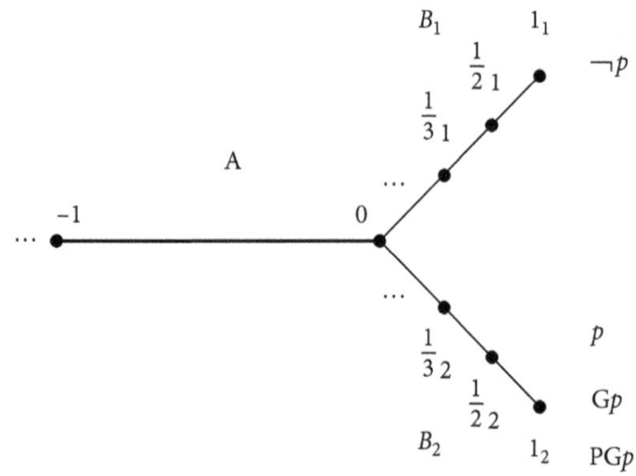

Within this structure, premiss P_1 is true despite branching. Each moment here has its immediate preceding moment; however, the last moment before the branching — moment 0 — is not an immediate predecessor of any other moment. Thus, if formula $p \wedge \mathsf{G}p$ is true at moment 1_2 (or at any moment preceding it), then there exists such immediately preceding it moment b that for any x if $b < x$, then formula p is true at x, for b belongs to only one branch within the branching.

Due to the branching of the presented model, the assumption of premiss P_1 does not exclude the branching of time. Consequently, it seems that the reconstruction by Arthur Prior's idea allows to consider indeterministic approach to the future. It is also worth to mention certain controversies related to time functor F.

Uttering a sentence:

(a) *Tomorrow it will rain.*

we state that no matter *which tomorrow* will be tomorrow (that is, regardless of which of the possible futures will occur), the rain will fall. Therefore, the literal reading of operator F corresponds to its Peircean understanding, and in this case, the obtained definition of possibility seems to assert the determinism: *a sentence is possible iff it is true or will be true.* On the other hand — as presented above — time does not need to be linear. How, then, to accommodate this apparent inconsistency?

The literal reading of operator F, called Peircean, requires the sentence preceded by operator F to be true at a certain point in each future branch. In logic $\mathbf{K}_t$ and in Prior's reconstruction a standard understanding of operator F is employed; it requires the sentence preceded by this operator to be true at least at one moment of the future. However, this reading is inconsistent with common intuitions regarding grammatical future tense. Thus understood operator F rather states that: *it may be the case that* ... then: *it will be the case that* .... Hence, the standard reading of operator F leads to the non-intuitive and, as a matter of fact, naive consequences — for we think about a common understanding of the grammatical future tense, whereas in the semantics a different interpretation is employed. Notwithstanding, if we wanted to employ in the semantics the Peircean literal reading of operator F, then, with the logical measures employed, the future tense would have had to be linear[212].

Nevertheless, continuing with regular tense interpretation of the time operators, Arthur Prior's reconstruction requires the time structure not to have the first moment, and each moment to have its immediate predecessor. As seen above, it does not have to mean the linearity of the time structure. For the time can be branching. Thus, the alternative futures are possible.

Employing the premisses formulated by Prior, one can consider one more version of the reconstruction of the Reasoning of Diodorus, which has been

212 See [120], pp. 93–95.

presented by Morton White[213]. It differs in certain details, including a weaker definition of possibility. This variant employs, however, axiom $\mathsf{F}A \wedge \mathsf{F}B \rightarrow \mathsf{F}(A \wedge B) \vee \mathsf{F}(A \wedge \mathsf{F}B) \vee \mathsf{F}(\mathsf{F}A \wedge B)$ that, in turn, excludes the time branching into the future.

9.2 Reconstruction of P. Øhrstrøm

The reconstruction proffered by P. Øhstrøm is of a preformal nature. Notwithstanding, it is in line with many intuitions of Benson Mates implicit in the presented above historical work on Stoic logic. This reconstruction can be found in the article *A new reconstruction of the master argument of Diodorus Cronus* [204] and also in the cited above work *Temporal Logic. From Ancient Ideas to Artificial Intelligence* [205]. Since the author does not make a proper distinction into syntax and semantics of the language in which he conducts his reasoning, it is difficult to carry out the reconstruction in a strictly formal manner, without altering the author's intentions. Therefore, let us present it in a form provided by its author, but with few modifications that will not impact his idea significantly.

9.2.1 Preliminaries

Assume that the sentences that **RDC** states about, are the propositional functions of arguments from set $\langle T, \leq \rangle$, that is, the sentences temporally indeterminate $\mathbf{S}_{\mathbf{TU}}$. Thus, take set of variables **Var** and for any $p_i \in \mathbf{Var}$ define function $p_i : T \longrightarrow \{0,1\}$. In addition to the sentential variables, we shall make use of the classical logical connectives. $\neg, \wedge, \vee, \rightarrow, \leftrightarrow, \forall, \exists$ and the equals sign $=$.

The following expressions can be the formulas: $p_i(t_j) = 1$, $p_i(t_j) = 0$, $t_i \leq t_j$, $t_i < t_j$ etc. Variables $t_1, t_2, \ldots$ will be ranging over set T, whereas constant n denotes an arbitrary but invariant moment called a present moment. Moreover, by default, if: A and B are the formulas, then also expressions, structured from them by means of the classical functors, are formulas.

In the language, there are also uninterpreted operators $\Box$, $\Diamond$ that correspond to the necessity and possibility of the Megarian. Øhstrøm assumes that an expression in a form of $A = 1$, where A is a predicate or another further specified expression that may occur in this context, is logically equivalent to expression $\neg A = 0$.

---

213 See [284], p. 129.

Furthermore, we shall define certain new connectives, in particular, Diodorus' implication:

$$DEF^{\to^D}: \; p_i \to^D q_j \equiv \forall_{t_1} \left(p_i(t_1) = 1 \to q_j(t_1) = 1 \right).$$

This definition expresses Diodorus' intuition that one sentence (propositional function) implicates another sentence (propositional function) iff for any values of the time parameter an appropriate material implication is true. Furthermore, Øhstrøm introduces a certain special operator defined as follows. For any t_1:

$$\mathsf{P} p_i(t_1) = 1 \Leftrightarrow \exists_{t_2} \left(t_2 < t_1 \wedge p_i(t_2) = 1 \right).$$

This operator, in relation to the sentence it acts on, should be read: *it has been the case that p_i is true*. In the given notation, it is stated that there is such moment prior to t_1, to which p_i assigns a value one.

There is also needed such propositional function q_0 that for any $t_1 \in T$ takes the following values:

$$q_0(t_1) = 1, \text{ if } t_1 = n - 1$$
$$q_0(t_1) = 0, \text{ in all other cases.}$$

The fact of the existence of such sentence q_0 that is true at a moment earlier by one than n and false at the other moments, we shall call premiss P_1.

According to the idea of Øhstrøm, Diodorus' premisses take the following form:

$$D1: \; \mathsf{P} p_i(n) = 1 \to \Box \mathsf{P} p_i(n) = 1$$
$$D2: \; \left(\Diamond p_i(t_1) = 1 \wedge p_i \to^D q_j \right) \to \Diamond q_j(t_1) = 1$$
$$D3: \; \Diamond p_0(n) = 1 \wedge \left(n \le t_1 \to p_0(t_1) = 0 \right).$$

We shall also make use of the following rule which states that certain modal expressions are equivalent to other modal expressions. It corresponds to Aristotle's principle.

$$P_2: \; \Box p_i(t_1) \leftrightarrow \neg \Diamond \neg p_i(t_1).$$

Moreover, in order to adequately express Diodorus' definition, there is the following premiss necessary:

$$P_3: \; \exists_{t_2} \left(t_1 \le t_2 \wedge p_i(t_2) = 1 \right) \to \Diamond p_i(t_1) = 1.$$

9.2.2 Reasoning

The reasoning by means of the presented tools is carried out as follows.

(1) From premiss *D*3, we know that there is some propositional function p_0 such that $\Diamond p_0(n) = 1$, but $\forall_{t_1}(n \leq t_1 \rightarrow p_0(t_1) = 0)$.
(2) Due to P_1 and the definition of operator P, we get that for any t_1: $\mathsf{P}q_0(t_1) = 1$ if $n \leq t_1$, and $\mathsf{P}q_0(t_1) = 0$ for all other cases. Informally, it means that $\neg\mathsf{P}q_0(t_1) = 0$ if $n \leq t_1$, and $\neg\mathsf{P}q_0(t_1) = 1$ for all other cases.
(3) We know that $\forall_{t_1}(n \leq t_1 \rightarrow p_0(t_1) = 0)$ and also from (2) that $\forall_{t_1}(n \leq t_1 \rightarrow \neg\mathsf{P}q_0(t_1) = 0)$. Whereas at the other moments $\neg\mathsf{P}q_0(t_1) = 1$. Accordingly $\forall_{t_1}(p(t_1) = 1 \rightarrow \neg\mathsf{P}q_0(t_1) = 1)$.
(4) Øhstrøm employs the above fact and $DEF^{\rightarrow^D}$ in order to proceed to the statement that $p_0 \rightarrow^D \neg\mathsf{P}q_0$. This transition is quite informal, for Diodorus' definition of implication relates to the sentential variables, not to the complex formulas.
(5) Taking advantage of the fact that $\Diamond p_0(n) = 1$, the above conclusion and *D*2, we can proceed to the statement that $\Diamond\neg\mathsf{P}q_0(n) = 1$.
(6) From definition P_2 it means that $\neg\Box\mathsf{P}q_0(n) = 1$, which is also an informal transition.
(7) However, since $\mathsf{P}q_0(n) = 1$, then by the virtue of *D*1, we get that: $\Box\mathsf{P}q_0(n) = 1$, what negates the conclusion from (6).

9.2.3 Definitions of modality

There has nothing been assumed about predicate p_0 and about n. The whole reasoning can be, thus, generalised into any function and any moment. Rejecting premiss *D*3 and accepting its negation, by the virtue of **CPL**, **FOL**, results in the following statement:

$$D\Diamond': \Diamond p_i(t_1) = 1 \rightarrow \exists_{t_2}(t_1 \leq t_2 \wedge p_i(t_2) = 1)$$

which is to express one side of Diodorus' definition. Composing it with P_3 results in a statement that expresses the complete content of the definition of possibility:

$$D\Diamond: \Diamond p_i(t_1) = 1 \leftrightarrow \exists_{t_2}(t_1 \leq t_2 \wedge p_i(t_2) = 1).$$

On the other hand, employing the laws of **CPL**, **FOL**, and P_2 — assuming that in the previous formula we can substitute $p_i(t_j) = 1$ with $\neg p_i(t_j) = 1$ — we get a statement that expresses the content of Diodorus' definition of necessity:

$$D\Box: \Box p_i(t_1) = 1 \leftrightarrow \forall_{t_2}(t_1 \leq t_2 \rightarrow p_i(t_2) = 1).$$

The issue of expanding the above definitions into arbitrary formulas would have required proper defining of the language and the logical tools employed in the reconstruction. The presented reconstruction is, however, of a preformal nature and its author has not defined the logical framework in which it had been carried out.

9.2.4 Time structure

The presented idea imposes certain restrictions on time structure $\langle T, \leq \rangle$. Since we consider a moment earlier than n, the time cannot have a beginning. The considered moment must be immediately preceding n, thus the time structure must also meet the condition of left-discreteness. In the presented reasoning, the proof steps no. 2 and 3 seem to exclude the branching of time in the future. Nevertheless, we can define the time structure that is branching and does not conflict with the proof steps no. 2 and 3. The structure in question is the one presented in the subchapter 9.1.4, within which the moment at which the branching occurs is not immediately preceding any other moment.

Another issue is the question of premisses $D1$ and $D2$, and carrying out a reduction by means of definitions $D\Diamond$ and $D\Box$. Due to employing the preformal intuitive tools in the theory, we must also rely on a somewhat vague procedure that results in the observation that premisses $D1$ and $D2$ do not require any additional assumptions in relation to the time structure.

Worrying is, though, the fact of employing premiss P_1, that is, the assumption that for any moment immediately preceding n there is such sentence from $\mathbf{S}_{\mathrm{TU}}$ that at this moment is true, and at any other points is false. This is an extralogical assumption.

The discussed reconstruction would have required structuring a proper logical tool. Then the conclusions regarding the issues of the time structure would have been firmer.

Conclusion

In the present work, we have attempted to outline the philosophical and logical framework that would enable us to answer the titular problem. We have developed a certain concept of time, arguing that by its means, to a certain extent, we can interpret various statements, the time-dependent ones as well as those whose content does not depend on time.

The presented concept of time meets various requirements imposed on the time understood not only in common or physical terms, but it also allows to consider such models of time that contradict these intuitions, in particular, the models of the branching time.

By means of the notion of the point time one can use various logical tools that are employed in the reasonings that consider the time parameter. We have presented the foundations of the new approach to the positional logic with operator R_i, which employs the structures of the coordinated time that is represented by special subsets of real numbers. However, the notion of the point time also corresponds to conventionally understood operator R_t and to the interpretation of the tense logic. Therefore, on the basis of one notion, it has been possible to interpret various approaches to the reconstruction of the Reasoning of Diodorus Cronus.

The overview of the reconstructions of the reasoning, in respect of the logical time structures entangled in it, suggests that it is fundamental for its purposes to postulate a linear time structure in its different variants: the dense or discrete structure, with no end or no beginning. In principle, one can thus assume that the original Diodorus' reasoning could had favoured the linear time, leading to certain deterministic consequences by means of the exclusion of the alternative futures. However, from today's point of view, it is not a necessary assumption. For some of the reconstructions allow the branching of time. In the last of the presented reconstructions, there occurs a possibility of some very special branches of time. However, in Prior's reconstruction, the matter involves a change in the meaning of the future operator. On the other hand, the last reconstruction is merely of an informal nature, and besides, it is based on certain additional extralogical assumptions.

During the preparation of this study, there occurred many additional problems that, even though associated to the study, reach beyond the scope of this research. Nevertheless, they are worth further analysis and developing by the logical means. Some of these issues have been suggested in the course of the study. I shall try to review and complete the most important ones.

In terms of time, one should tackle the issue of the interval time, that is, the time seen as a set of linearly ordered sets of points, and the time without points, that is, close to the phenomenological approach to time. In order to obtain the possibility to apply both the issues to the problem of the reasoning of Diodorus and to the similar ones, a separate, quite extensive research is required.

Another issue occurring in the study is a correct and full description of the logical systems that arise within the proposed interpretation of operator R_i. Specifically important and interesting is a question about their satisfying axiomatisation that distinguishes certain time models, and about metatheorems associated with it, e.g. the completeness theorem. The issue of these operators also involves a question of which time structures are the canonical structures, that is, which can be coordinated (e.g. whether each continuous structure can be coordinated).

Thus, by engaging in a certain problem and outlining its background, this work has generated some new interrelated issues. Therefore, the presentation of certain propositions of an approach and a solution to the problem of the logical time structures within the Reasoning of Diodorus Cronus constitutes a starting point and a background for the development and attempts to solve the issues suggested herein.

Summary

This book is devoted to the old problem of Master Argument. The argument was made by an ancient logician and philosopher Diodorus Cronus. Unfortunately, the famous Master Argument is a lost reasoning of Diodorus Cronus.

As ancient commentators claimed, Diodorus had supported the standpoint that: *something is possible iff it is or will be true* and *something is necessary iff it is and will always be true*. In order to justify this view, Diodorus assumed the negation of the first sentence and showed that it provides a contradiction. The premises of his reasoning were the following sentences:

(*D*1) *Every proposition true about the past is necessary.*
(*D*2) *An impossible proposition cannot follow from a possible one.*
(*D*3) *There is a proposition which is possible, but which neither is nor will be true.*

Because Diodorus thought that the first two premises were plausible, he denied the last one, obtaining his definition of the possibility. Using the Aristotle's law: *it is not possible that it is not p iff it is necessary that p*, he could also obtain the definition of the necessity.

In this book we look at the Master Argument from the perspective of so called Sea Battle Tomorrow, an argument made by Aristotle to support the indeterministic view. In consequence we treat Master Argument as an argument for determinism.

The book contains three parts. In the first part of the book, in three chapters, we outline relationships between three fundamental notions: time, truth, in/determinism:

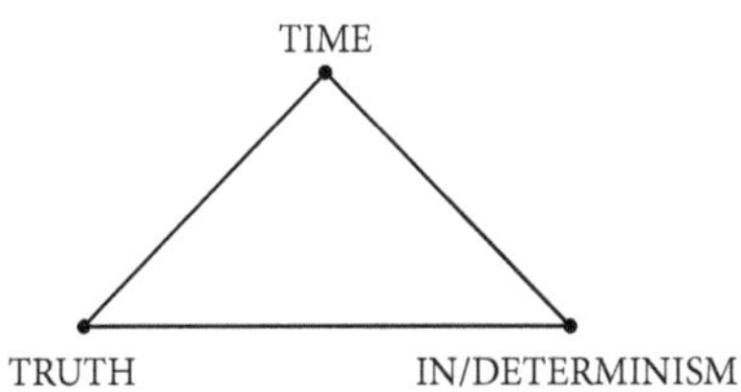

In the second part, in chapters: *The problem*, *Dates, times, propositions vs. time's structures* and *Logic*, historical evidence of the discussion between Diodorus and Aristotle are described, and logical tools for further reconstructions of the Master Argument are prepared.

In the last part — *Solutions* — we present and discuss various attempts of reconstructions of the Master Argument. They can be divided into two groups, reconstructions that based on:

1. reconstructions that based on $\mathcal{R}$-operator, chronological logics (originated by Jerzy Łoś), e.g., the reconstruction of Nicholas Rescher
2. others, not located within any logical frame, including tense logics.

The main aim of the book is to examine, in a strictly logical way, the reconstructions of the lost argument. We want to know whether it is possible to place Diodorus argument within a frame of logic that allows branching time. In other words, we want to know whether Diodorus modal notions are consistent with any of the indeterministic, logical frames of time. Finally, we demonstrate that it is possible.

Bibliography

[1] Ajdukiewicz K., *Czas* (*Time*), (in:) *Język i poznanie* (*Language and Cognition*), vol. 2, Warszawa, PWN, 1985, pp. 384–387.

[2] Ajdukiewicz K., *Język i znaczenie* (*Language and meaning*), (in:) *Język i poznanie* (*Language and Cognition*), vol. 2, Warszawa, PWN, 1985, pp. 145–174.

[3] Ajdukiewicz K., *Kategorie syntaktyczne i antynomie logiczne* (*Syntactical categories and logical antinomies*), Filozofia Nauki, 1 (1993), pp. 163–182.

[4] Ajdukiewicz K., *O spójności syntaktycznej* (*On syntactic consistency*), (in:) *Język i poznanie* (*Language and Cognition*), vol. 1, Warszawa, PWN, 1985, pp. 222–242.

[5] Ajdukiewicz K., *O znaczeniu wyrażeń* (*On the meaning of expressions*), (in:) *Język i poznanie* (*Language and Cognition*), vol. 2, Warszawa, PWN, 1985, pp. 102–136.

[6] Ajdukiewicz K., *Proposition as the connotation of sentence*, Studia Logica, 20 (1967), pp. 87–98.

[7] Ajdukiewicz K., *Pragmatic Logic*, Dordrecht, Reidel, 1974.

[8] Ajdukiewicz K., *Związki składniowe między członami zdań oznajmujących* (*Syntactical relations between members of declarative sentences*), (in:) *Język i poznanie* (*Language and Cognition*), vol. 2, Warszawa, PWN, 1985, pp. 341–364.

[9] Allen J. F., *Time and time again: the many ways to represent time*, International Journal of Intelligent Systems, 6 (1991), pp. 341–355.

[10] Amsterdamski S., *O różnych pojęciach determinizmu* (*On different notions of determinism*), (in:) Z. Horbaczewski (ed.), *Główne zagadnienia filozofii* (*Main Issues of Philosophy*), Warszawa, PWN, 1967, pp. 204–214.

[11] Aristotle, *On Interpretation*, E. M. Edghill (trans.), (in:) W. D. Ross and J. A. Smith (eds.), *The Works of Aristotle*, vol. 1, London, Oxford University Press, 1996, pp. 46–79.

[12] Aristotle, *Physics*, R. Hope (trans.), Lincoln, University of Nebraska Press, 1961.

[13] Aristotle, *Prior Analytics*, A. J. Jenkinson (trans.), (in:) W. D. Ross and J. A. Smith (eds.), *The Works of Aristotle*, vol. 1, London, Oxford University Press, 1996, pp. 80–243.

[14] Aristotle, *Metaphysics*, W. D. Ross (trans.), (in:) W. D. Ross and J. A. Smith (eds.), *The Works of Aristotle*, vol. 8, London, Oxford University Press, 1996.

[15] Arsenijević M., *Determinism, indeterminism and flow of time*, Erkenntnis, 56 (2002), pp. 123–150.

[16] Augustine of Hippo, *Confession*, F. J. Sheed (trans.), M. P. Foley (ed.), Indianapolis/Cambridge, Hackett Publishing Company, 2006.

[17] Augustynek Z., *Natura czasu* (*Nature of Time*), Warszawa, PWN, 1975.

[18] Augustynek Z., *Time. Past, Present, Future*, Dordrecht/Boston/London, Kluwer Academic Publishers, 1991.

[19] Augustynek Z., *Własności czasu* (*Properties of Time*), Warszawa, PWN, 1970.

[20] Aveni A., *Empires of Time: Calendars, Clocks and Cultures*, London, Tauris Parke Paperbacks, 2000.

[21] Bar-Hillel Y., *Indexical expressions*, Mind, 63 (1954), pp. 359–379.

[22] Barnes J., *Introduction*, (in:) K. Algra, J. Barnes, J. Mansfeld and M. Schofield (eds.), *The Cambridge History of Philosophy*, Cambridge, Cambridge University Press, 1999, pp. 65–76.

[23] Barnes J., *Logic*, (in:) K. Algra, J. Barnes, J. Mansfeld and M. Schofield (eds.), *The Cambridge History of Philosophy*, Cambridge, Cambridge University Press, 1999, pp. 77–176.

[24] Barwise J. and J. Etchemendy, *The Liar. An Essay on Truth and Circularity*, New York, Oxford University Press, 1987.

[25] Barwise J. and J. Perry, *Situations and Attitudes*, Cambridge, MIT Press, 1983.

[26] Bäuerle R., *Tense logics and natural language*, Synthese, 40 (1979), pp. 225–230.

[27] Bäumer B., *Empiryczne pojmowanie czasu*, K. Byrski (trans.), (in:) A. Zajączkowski (ed.), *Czas w kulturze*, Warszawa, PIW, 1988, pp. 139–149. Eng. version: *Empirical apperception of time: Kala and Karman. Time and history in the Indian tradition*, (in:) L. Gardet (ed.), *Cultures and Time*, Paris, The Unesco Press, 1976.

[28] Beaver D., *Pressuposition*, (in:) J. van Benthem and A. ter Meulen (eds.), *Handbook of Logic and Language*, Cambridge, MIT Press, 1997, pp. 939–1008.

[29] Benthem J. F. K. K., *Logic of Time*, Dordrecht, D. Reidel Publishing Company, 1983.

[30] Bennett M., *A guide to the logic of tense and aspect in english*, Logique and Analyse, 20 (1977), pp. 491–515.

[31] Bergson H., *Matter and Memory*, N. M. Paul and W. S. Palmer (trans.), New York, Cosimo, 2007.

[32] Bielecki M. W., *Czas i stawanie się w sztucznych i naturalnych układach kognitywnych (Time and becoming in aritificial and natural cognitive systems)*, Filozofia Nauki, 4 (1995), pp. 21–32.

[33] Black M., *The semantic definition of truth*, Analysis, 8, 4 (1948), pp. 49–63.

[34] Bohannan P., *Concepts of time among the Tiv of Nigeria*, Southwestern Journal of Anthropology, 9 (1953), pp. 251–262.

[35] Boltzmann L., *Vorlesungen über Gastheorie*, Leipzig, Barth, 1896.

[36] Borkowski L., *Pewna wersja definicji klasycznego pojęcia prawdy (Certain version of definition of classical notion of truth)*, Roczniki Filozoficzne, 1 (1980), pp. 119–131.

[37] Brentano F., *Psychology from an Empirical Standpoint*, A. C. Rancurello, D. B. Terrell and L. McAlister (trans.), London, Routledge, 1973.

[38] Brentano F., *Wahrheit und Evidenz*, Leipzig, F. Meiner, 1930.

[39] Broad C. D., *Time*, (in:) J. Hastings (ed.), *Encyclopedia of Religion and Ethics*, vol. 12, Edinburgh, Clark, 1921.

[40] Bunge M., *Dictionary of Philosophy*, Amherst/New York, Prometheus Books, 1999.

[41] Bunge M., *The Myth of Simplicity*, New York, Prentice Hall, 1963.

[42] Bunge M., *Causality and Modern Science*, New York, Dover Publications, 1979.

[43] Bunge M., *Philosophy of Science*, vols. 1 and 2, New Brunswick/London, Transaction Publishers, 1998.

[44] Burgess J. P., *Basic tense logic*, (in:) D. Gabbay and F. Guenthner (eds.), *Handbook of Philosophical Logic*, vol. 2, Dordrecht/Boston/Lancaster, D. Reidel Publishing Company, 1984, pp. 89–133.

[45] Carnap R., (in:) M. Gardner (ed.), *An Introduction to the Philosophy of Science*, New York, Dover Publications, 1995.

[46] Carnap R., *Meaning and Necessity*, Chicago, University of Chicago Press, 1956.

[47] Carnap R., *Testability and meaning*, Philosophy of Science, 3 (1936), pp. 419–471.

[48] Carnap R., *Über die Abhängigkeit der Eigenschaften des Raumes von denen der Zeit*, Kantstudien, 30 (1925), pp. 331–345.

[49] Cocburn D., *Other Times. Philosophical Perspectives on Past, Present and Future*, Cambridge, Cambridge University Press, 1997.

[50] Cresswell M. J., *Logics and Languages*, London, Methuen and Co. Ltd., 1973.

[51] Crombie A. C., *Medieval and Early Modern Science*, vol. 1, Cambridge, Harvard University Press, 1963.

[52] Church A., *Introduction to Mathematical Logic*, vol. 1, Princeton, Princeton University Press, 1956.

[53] Cover J. A. and G. A. Hartz, *Space and time in the Leibnizian metaphysics*, Nôus, 22 (1986), pp. 493–519.

[54] Czeżowski T., *Nazwy okazjonalne oraz imiona własne* (*Occasional names and proper nouns*), (in:) *Odczyty filozoficzne* (*Philosophical Readings*), Toruń, Towarzystwo Naukowe w Toruńiu, 1958, pp. 148–159.

[55] Damasio A. R., *Remembering when*, Scientific American, 287, 3 (2002), pp. 66–73.

[56] Dąmbska I., *Z filozofii imion własnych* (*On philosophy of proper nouns*), Kwartalnik Filozoficzny, 18 (1949), pp. 241–261.

[57] Descartes R., *Meditations on First Philosophy: With Selections from the Objections and Replies*, J. Cottingham (trans.), Cambridge, Cambridge University Press, 1996.

[58] Dewey J., *Logic. The Theory of Inquiry*, New York, H. Holt and Co., 1938.

[59] Diels H., *Die Fragmente der Vorsokratiker*, Berlin, 1903.

[60] Diogenes Laertius, *Lives and Opinions of Eminent Philosophers*, C. D. Yonge (trans.), CreateSpace Independent Publishing Platform, 2014.

[61] Doliński D. and W. Łukaszewski, *Typy motywacji* (*Types of motivation*), (in:) J. Strelau (ed.), *Psychologia. Podręcznik akademicki* (*Psychology. Academic Handbook*), vol. 2, Gdańsk, GWP, 2000, pp. 468–491.

[62] Doob L., *Time: cultural and social aspects*, (in:) T. Carlstein, D. Parkes and N. Trift (eds.), *Making Sense of Time*, vol. 1, London, Edward Arnold, 1978, pp. 56–65.

[63] Dohrn-van Rossum G., *History of the Hour: Clocks and Modern Temporal Orders*, Th. Dulnap (trans.), Chicago, University of Chicago Press, 1996.

[64] Döring K., *Die Megariker. Kommentierte Sammlung der Testimonien*, Amsterdam, Verlag B. R. Grüner N. V., 1972.

[65] Duda R., *Czas jako archetyp w matematyce* (*Time as an archetype in mathematics*), (in:) M. Heller and J. Mączka (eds.), *Czas…* (*Time…*), Kraków, Biblos, 2001, pp. 11–18.

[66] Dummett M., *The reality of the past*, Proceedings of the Aristotelian Society, 69 (1969), pp. 239–258.

[67] Earman J., *A Primer on Determinism*, Dordrecht, Reidel Publishing Company, 1986.

[68] Einstein A., *The Meaning of Relativity. Four Lectures Delivered at Princeton University, May, 1921*, Princeton, Princeton University Press, 1923.

[69] Epictetus, *The Discourses*, G. Long (trans.), London, George Bell and Sons, 1903.

[70] Evans-Pritchard E. E., *The Nuer*, Oxford, Clarendon Press, 1940.

[71] Faye J., *The Reality of the Future. An Essay on Time, Causation and Backward Causation*, Odense, Odense University Press, 1989.

[72] Faye J., U. Scheffler and M. Urchs, *Introduction*, (in:) J. Faye, U. Scheffler and M. Urchs (eds.), *Perspectives on Time*, Dordrecht/Boston/London, Kluwer Academic Publishers, 1997, pp. 1–58.

[73] Fitch F. B., *A logical analysis of some value concepts*, Journal of Symbolic Logic, 28 (1963), pp. 135–142.

[74] Fitting M. and R. L. Mendelsohn, *First-Order Modal Logic*, Dorderecht, Kluwer, 1998.

[75] Frege G., *Fragmenty z Grundlagen der Arithmetik*, B. Wolniewicz (trans.) (in:) *Pisma semantyczne (Semantical writings)*, Warszawa, PWN, 1977, pp. 10–17. Eng. version: *Translation of parts of Frege's Grundgesetze der Arithmetik*, M. Black (trans.), (in) P. Geach and M. Black (eds.), *Philosophical Writings of Gottlob Frege*, Oxford, Blackwell, 1980.

[76] Frege G., *Sens i znaczenie*, B. Wolniewicz (trans.) (in:) *Pisma semantyczne (Semantical writings)*, Warszawa, PWN, 1977, pp. 60–88. Eng. version: *On sense and reference*, M. Black (trans.), (in) P. Geach and M. Black (eds.), *Philosophical Writings of Gottlob Frege*, Oxford, Blackwell, 1980.

[77] Frege G., *The Thought: A Logical Inquiry*, Mind, 65, 259 (1956), pp. 289–311.

[78] Furley D., *Cosmology*, (in:) K. Algra, J. Barnes, J. Mansfeld and M. Schofield (eds.), *The Cambridge History of Philosophy*, Cambridge, Cambridge University Press, 1999, pp. 412–451.

[79] Gardet L., *Poglądy Muzłumanów na czas i historię. Rozważania z zakresu typologii kultur*, B. R. Zagórski (trans.), (in:) A. Zajączkowski (ed.), *Czas w kulturze*, Warszawa, PIW, 1988, pp. 151–180. Eng. version: *Muslim views on time and history (An essay in cultural typology)*, (in:) L. Gardet (ed.), *Cultures and Time*, Paris, Unesco Press, 1976.

[80] Gaskin R., *The Sea Battle and the Master Argument: Aristotle and Diodorus Cronus on the Metaphysics of the Future*, Berlin, De Gruyter, 1995.

[81] Garson J. and N. Rescher, *Topological logic*, Journal of Symbolic Logic, 33 (1968), pp. 537–548.

[82] Gibbs J. W., *Elementary Principles of Statistical Mechanics*, London, Edward Arnorld, 1902.

[83] Gleick, *Faster: The Acceleration of Just About Everything*, New York, Vintage Books, 1999.

[84] Gołosz J., *Spór o naturę czasu i przestrzeni* (*Dispute over the Nature of Time and Space*), Kraków, Wydawnictwo UJ, 2001.

[85] Gorzka C., *Przyczynek do zagadnienia definiowalności momentów czasu jako zbiorów zdarzeń* (*Input to the issue of definitability time moments as sets of occurances*), (in:) J. Perzanowski, A. Pietruszczak and C. Gorzka (eds.), *Filozofia/Logika. Filozofia logiczna*, Toruń, Wydawnictwo UMK, pp. 115–129.

[86] Grayling A. C., *An Introduction to Philosophical Logic*, Oxford, Blackwell Publishers, 1997.

[87] Griffin D. R., *Introduction: time and the fallacy misplaced concreteness*, (in:) D. R. Griffin (ed.), *Physics and the Ultimate Significance of Time*, Albany, State University of New York Press, 1986, pp. 1–48.

[88] Grobler A., *Prawda a względność* (*Truth vs. Relativity*), Kraków, Aureus, 2000.

[89] Grünbaum A., *Carnap's views on the foundations of geometry*, (in:) P. A. Schlipp (ed.), *The Philosophy of Rudolf Carnap*, La Salle, The Open Court Publishing Co., 1962, pp. 599–684.

[90] Grünbaum A., *Temporally-asymmetric principles parity between explanation and prediction, mechanism versus teleology*, Philosophy of Science, 29 (1962), pp. 146–170.

[91] Grünbaum A, *Popper on irreversibility, to appear in the critical approach*, (in:) M. Bunge (ed.), *Essays in Honour of Karl Popper*, Glencoe, Free Press, 1962, pp. 142–167.

[92] Grzegorczyk A., *Mała propedeutyka filozofii naukowej* (*Little Propaedeutics to Scientific Philosphy*), Warszawa, Instytut Wydawniczy Pax, 1989.

[93] Gutowski P. and T. Szubka, *Wprowadzenie* (*Introduction*), (in:) P. Gutowski and T. Szubka (eds.), *Filozofia brytyjska u schyłku XX wieku* (*British Philosophy at the End of XXth Century*), Lublin, Towarzystwo Naukowe KUL, 1998, pp. 57–65.

[94] Hajnicz E., *Reprezentacja logiczna wiedzy zmieniającej się w czasie*, (*Logical Representation of Knowledge Changing in Time*), Warszawa, Akademicka Oficyna Wydawnicza PLJ, 1996.

[95] Haldane J., *Mind-world identity and the anti-realist challenge*, (in:) J. Haldane and C. Wright (eds.), *Reality, Representation, and Projection*, New York, Oxford University Press, 1993, pp. 15–37.

[96] Hall E. T., *The Dance of Life: The Other Dimension of Time*, Garden City, Anchor Press/Doubleday, 1983.

[97] Hankinson R. J., *Determinism and indeterminism*, (in:) K. Algra, J. Barnes, J. Mansfeld and M. Schofield (eds.), *The Cambridge History of Philosophy*, Cambridge, Cambridge University Press, 1999, pp. 513–541.

[98] Hart W. D. and C. McGinn, *On propositions*, Notre Dame Journal of Formal Logic, 19 (1978), pp. 299–306.

[99] Heller M., *Filozofia świata* (*Philosophy of the World*), Kraków, Wydawnictwo Znak, 1992.

[100] Heller M., M. Lubański and Sz. W. Ślaga, *Zagadnienia filozoficzne współczesnej nauki. Wstęp do filozofii przyrody* (*Philosophical Issues in Contemporary Science. Introduction to Philosophy of Nature*), Warszawa, Akademia Teologii Katolickiej, 1982.

[101] Heller M. and D. J. Raine, *The Science of Space-Time*, Tuscan, Pachart, 1981.

[102] Heller M. and A. Staruszkiewicz, *A physicist's view on the polemics between Leibniz and Clarke*, Organon, 11 (1975), pp. 205–213.

[103] Hempel C. G. and P. Oppenheim, *Studies of the logic of explanation*, Philosophy of Science, 15 (1948), pp. 135–178.

[104] Hintikka J., *Time and Necessity. Studies in Aristotle's Theory of Modality*, London, Oxford University Press, 1973.

[105] Hintikka J., *Aristotle and the 'Master Argument' of diodorus*, American Philosophical Quarterly, 1 (1964), pp. 101–114.

[106] Higginbotham J., *Tense, indexicality and consequence*, (in:) J. Butterfield (ed.), *The Arguments of Time*, Oxford, The British Academy, 1999, pp. 197–215.

[107] Higginbotham J., *Why is sequence of tense obligatory?*, (in:) G. Preyer and G. Peter (eds.), *Logical Form and Language*, Oxford, Clarendon Press, 2002, pp. 207–227.

[108] Holbach P. H., *The System of Nature, or, Laws of the Moral and Physical World*, H. D. Robinson (trans.), Boston, J. P. Mendum, 1889.

[109] Honderich T., *How Free Are You?: The Determinism Problem*, Oxford, Oxford University Press, 2002.

[110] Horwich P., *Asymmetries in Time. Problems in the Philosophy of Science*, Cambridge, MIT Press, 1987.

[111] Hume D., *Enquiries Concerning the Human Understanding and Concerning the Principles of Morals*, Oxford, Oxford University Press, 1902.

[112] Husserl E., *Logical Investigations*, vols. 2.1 and 2.2, J. N. Findlay (trans.), D. Moran (ed.), London/New York, Routledge, 2001.

[113] Husserl E., *Cartesian Meditations. An Introduction to Phenomenology*, D. Cairns (trans.), Seventh impression, The Hague/Boston/London, Martiuns Nijhoff Publishers, 1982.

[114] Husserl E., *On the Phenomenology of the Consciousness of Internal Time (1983–1917)*, J. Brough (trans.), Dordrecht, Kluwer Academic Publishers, 1991.

[115] Ingarden R., *The Literary Work of Art*, G. G. Grabowicz (trans.), Evanston, Northwestern University Press, 1973.

[116] Jaeger W., *Paideia: The Ideals of Greek Culture*, vol. 1, G. Highet (trans.), Oxford, Oxford University Press, 1986.

[117] James W., *Pragmatism: A New Name for Some Old Ways of Thinking*, Cambridge, Harvard University Press, 1975.

[118] James W., *Principles of Psychology*, New York, Henry Holt, 1890.

[119] Jarmużek T. and A. Pietruszczak, *Completeness of minimal positional calculus*, Logic and Logical Philosophy, 13 (2004), pp. 147–162.

[120] Jarmużek T. and A. Pietruszczak, *The tense logic for master argument in prior's reconstruction*, Studia Logica, 2 (2009), pp. 85–108.

[121] Jarmużek T. and M. Tkaczyk, *Normalne logiki pozycyjne* (*Normal Pos*), Lublin, Wydawnictwo Naukowe KUL, 2015.

[122] Jarmużek T., *Rekonstrukcje Rozumowania Diodora Kronosa w ontologii czasu punktowego* (*Reasoning of Diodorus Cronus in pointwise time ontology*), Analiza i Egzystencja, 3 (2006), pp. 197–215.

[123] Jarmużek T., *Rozumowanie Diodora Kronosa a problem anizotropii czasu* (*Reasoning of Diodorus Cronus vs the problem of time anisotropy*), (in:) W. Krajewski and B. Szumilewicz (eds.), *Nauka, świat, człowiek* (*Science, World, Human*), volume dedicated to prof. Irena Szumilewicz, Gdańsk, Wydawnictwo UG, 2005, pp. 29–40.

[124] Jarmużek T., *Różne oblicza czasu i ich wspólny fundament* (*Different faces of time and their common foundation*), Przegląd filozoficzny. Nowa Seria, Warszawa, 3, 55 (2005), pp. 47–55.

[125] Jokić A. and Q. Smith (eds.), *Time, Tense and Reference*, Cambridge, MIT, 2003.

[126] Kant I., *Critique of Pure Reason*, W. S. Pluhar (trans.), Indianapolis/Cambridge, Hackett Publishing Company, 1996.

[127] Kałuszyńska E., *Inne oblicze realizmu* (*Different face of realism*), *(in:) U. M. ̓egleń (ed.), Pragmatyzm i filozofia Hilarego Putnama (Pragmatism and philosophy of Hilary Putnam*), Toruń, Wydawnictwo UMK, 2001, pp. 77–87.

[128] Kałuszyńska E., *Nasz czas* (*Our time*), (in:) M. Heller and J. Mączka (eds.), *Czas…* (*Time…*), Kraków, Biblos, 2001, pp. 141–154.

[129] Kałuszyńska E., *Teorie i układy deterministyczne* (*Deterministic theories and systems*), Studia Filozoficzne, 8 (1974), pp. 45–52.

[130] Kaplan D., *On the logic of demonstratives*, Journal of Philosophical Logic, 8 (1978), pp. 99–115.

[131] Katz R., *An introduction to the Special Theory of Relativity*, Princeton, Van Nostrand Company, 1964.

[132] Khan Ch., *The Verb "Be" in Ancient Greek*, Dordrecht, Reidel, 1973.

[133] Kneale M. and W. Kneale, *The Development of Logic*, Oxford, Clarendon Press, 1984.

[134] Krajewski W., *Determinizm i indeterminizm* (*Determinism and indeterminism*), (in:) *Filozofia a nauka. Zarys encyklopedyczny* (*Philosophy vs. Science. Encyclopedical Outline*), Wrocław, PAN, 1987, pp. 86–92.

[135] Krajewski W., *Four conceptions of causation*, (in:) W. Krajewski (ed.), *Polish Essays in the Philosophy of the Natural Sciences*, Dordrecht, Reidel, 1982, pp. 223–235.

[136] Krajewski W., *Konieczność , przypadek, prawo statystyczne* (*Necessity, Contingency, Statistical Law*), Warszawa, PWN, 1977.

[137] Krajewski W., *Związek przyczynowy* (*Causal Relation*), Warszawa, PWN, 1967.

[138] Kripke S. A., *A completness theorem in modal logic*, Journal of Symbolic Logic, 24 (1959), pp. 1–14.

[139] Kripke S. A., *Semantical analysis of modal logic I*, Zeitschrift für mathematische Logik und Grundlagen der Mathematik, 9 (1963), pp. 67–96.

[140] Kripke S. A., *Semantical considerations on modal logics*, Acta Philosophica Fennica, 16 (1963), pp. 83–94.

[141] Kuhn S., *The pragmatics of tense*, Synthese, 40 (1979), pp. 231–263.

[142] Künne W., *Prawda* (*Truth*), K. Krzemieniowa (trans. from German), (in:) E. Martensa and H. Schnädelbacha (eds.) *Filozofia. Podstawowe pytania* (*Philosophy. Elementary Questions*), Warszawa, Wiedza Powszechna, 1995, pp. 138–196.

[143] Künne W., *Putnama wariacje na temat prawdy* (*Putnam's variations on the issue of truth.*), A. Derra-Włochowicz (trans.), (in:) U. M. Żegleń (ed.), *Pragmatyzm i filozofia Hilarego Putnama* (*Hilary Putnam's Pragmatism and Philosophy*), Toruń, Wydawnictwo UMK, 2001, pp. 157–172.

[144] La Mettrie J. O., *Machine Man and Other Writings*, A. Thompson (trans. and ed.), Cambridge, Cambridge University Press, 1996.

[145] Laplace P. S., *Théorie analytique des probabilités*, Paris, V. Courcier, 1820.

[146] Larre C., *Empiryczne pojmowanie czasu a koncepcja historii w myśli chińskiej*, B. Chwedeńczuk (trans.), (in:) A. Zajączkowski (ed.), *Czas w kulturze*, Warszawa, PIW, 1988, pp. 69–110. Eng. version: *The empirical*

apperception of time and the conception of history in chinese thought, (in:) L. Gardet (ed.), *Cultures and Time*, Paris, Unesco Press, 1976.

[147] Lawrance N., *Levels of language of discourse about time*, (in:) J. T. Fraser, F. C. Haber and G. H. Müller (eds.), *The Study of Time*, vol. 3, Berlin, Springer-Verlag, 1978, pp. 22–49.

[148] Leach E. R., *Rethinking Antropology*, London, Athlone Press, 1961.

[149] Le Goff J., *Labor time in the 'crisis' of the fourteenth century: From medieval to modern time*, (in:) J. Le Goff (ed.), *Time, Work, and Culture in the Middle Ages*, Chicago, University of Chicago Press, pp. 43–52.

[150] Leibniz G. W. and S. Clarke, (in:) R. Ariew (ed.), *Correspondence*, Indianapolis/Cambridge, Hackett Publishing Company, 2000.

[151] Leibniz G. W., *Dialogue on human freedom and the origin of evil*, R. Ariew (trans.), (in:) R. Ariew and D. Garber (eds.), *Philosophical Essays*, Indianapolis, Hackett Publishing Company, 1989, pp. 111–117.

[152] Leibniz G.W., *Confessio philosophi, Papers Concerning the Problem of Evil 1671–1678*, R. C. Sleigh (trans. and ed.), New Haven/London, Yale University Press, 2005.

[153] Lemmon E. J., *Sentences, statements and propositions*, (in:) B. Williams and A. Montefiore (eds.), *British Analytical Philosophy*, London, Routledge and Paul Kegan, 1966, pp. 87–107.

[154] Lepore E. and K. Ludwig, *Outline for a truth-conditional semantics for tense*, (in:) A. Jokić and Q. Smith (eds.), *Time, Tense and Reference*, Cambridge, MIT, 2003, pp. 49–105.

[155] Le Poidevin R. and M. MacBeath, *Introduction*, (in:) R. Le Poidevin and M. MacBeath (eds.), *The Philosophy of Time*, Oxford/New York, Oxford University Press, 1997, pp. 1–20.

[156] Le Poidevin R. and D. H. Mellor, *Time, change, and the 'Indexical Fallacy'*, Mind, 4 (1987), pp. 534–538.

[157] Lewis D., *New work for a theory of universals*, Australiasian Journal of Philosophy, 63 (1983), pp. 343–377.

[158] Levinson S. C., *Pragmatics*, London, Cambridge University Press, 1983.

[159] Liddell H. G. and R. Scott, *A Greek English Lexicon*, Oxford, Clarendon Press, 1996.

[160] Lloyd G. E. R., B. Chwedeńczuk (trans.), (in:) A. Zajączkowski (ed.), *Czas w kulturze*, Warszawa, PIW, 1988, pp. 207–258. Eng. version: *Views on time in Greek thought*, (in:) L. Gardet (ed.), *Cultures and Time*, Paris, Unesco Press, 1976.

[161] Locke J., *An Essay Concerning Human Understanding*, vol. 2, Alexander Campbell Fraser (ed.), Oxford, Clarendon Press, 1894.

[162] Lowe E. J., *The indexical fallacy in McTaggart's proof of the unreality of time*, Mind, 1 (1987), pp. 62–70.

[163] Lowe E. J., *Replay to Le Poidevin and Mellor*, Mind, 4 (1987), pp. 539–542.

[164] Lycan W. G., *Philosophy of Language. A Contemporary Introduction*, London/New York, Routledge, 2000.

[165] Ładosz J., *Wielowartościowe rachunki zdań a rozwój logiki* (*Many-valued Propositional Calculi and Development of Logic*), Warszawa, Książka i Wiedza, 1961.

[166] Łoś J., *Podstawy analizy metodologicznej kanonów Milla* (*Foundations of the methodological analysis of Mill's canons*), Annales Universitatis Mariae Curie-Skłodowska, 2 (1947), pp. 269–301.

[167] Łukasiewicz J., *On determinism*, (in:) L. Borkowski (ed.), *Selected Works*, Amsterdam, North-Holland, 1970, pp. 110–128.

[168] Łukasiewicz J., *On three-valued logic* (in:) L. Borkowski (ed.), *Selected Works*, Amsterdam, North-Holland, 1970, pp. 87–88.

[169] Łukasiewicz J., *On the principle of the excluded middle*, History and Philosophy of Logic, 8 (1987), pp. 67–69.

[170] Łukasiewicz J., *Aristotle's Syllogistic from the Standpoint of Modern Formal Logic*, Oxford, Clarendon Press, 1957.

[171] Łukasiewicz J., *A system of modal logic*, (in:) L. Borkowski (ed.), *Selected Works*, Amsterdam, North-Holland, 1970, pp. 352–390.

[172] Majewski Z., *Mechanika klasyczna a zasada determinizmu* (*Classical mechanics and a principle of determinism*), Studia Filozoficzne, 11 (1988), pp. 15–23.

[173] Malebranche N., *De la recherche de la vérité*, Paris, J. Vrin, 1945.

[174] Malebranche N., *Entretiens sur la métaphysique et sur la religion*, Paris, A. Colin, 1922.

[175] Malinowski J., *The Pragmatic interpretation of utterances*, Logic and Logical Philosophy, 11/12 (2003), pp. 115–127.

[176] Maxwell J. C., *On the dynamical evidence of the molecular constitution of bodies*, Journal of the Chemical Society, 13 (1875), p. 493–508.

[177] Mates B., *Analytic sentences*, Philosophical Review, 60 (1951), pp. 525–534.

[178] Mates B., *Stoic Logic*, Berkeley, University of California Press, 1961.

[179] Mates B., *Diodorean implication*, Philosophical Review, 58 (1949), pp. 234–244.

[180] McArthur R. P., *Tense and temporally neutral paraphrase*, Australasian Journal of Philosophy, 1 (1974), pp. 28–36.

[181] McArthur R. P., *Tense Logic*, Dordrecht/Boston, D. Reidel, 1976.

[182] McDowell J., *On "the reality of past"*, (in:) C. Hookway and P. Pettit (eds.), *Action and Interpretation: Studies in Philosophy of Social Sciences*, Cambridge, Cambridge University Press, 1978, pp. 127–144.

[183] McMullin E., *Newton on Matter and Activity*, Notre Dame/London, University of Notre Dame, 1978.

[184] McTaggart J. M. E., *The unreality of time*, (in:) R. Le Poidevin and M. MacBeath (eds.), *The Philosophy of Time*, Oxford/New York, Oxford University Press, 1997, pp. 23–34.

[185] Mehlberg H., *Physical laws and time's arrow*, (in:) H. Feigl and G. Maxwell (eds.), *Current Issues in the Philosophy of Science*, New York, 1961, pp. 105–108.

[186] Mehlberg H., *Essai sur la théorie causale du temps*, Studia Philosophica, 1 (1935), pp. 119–260.

[187] Meinong A., *Über Annahmen*, (in:) R. Haller and R. Kindinger (eds.), *Alexius Meinong Gesamtausgabe*, vol. 4, Graz, Akademische Druck und Verlagsanstalt, 1977.

[188] Mellor D. H., *Real Time II*, London/New York, Routledge, 1998.

[189] Meyer J. J. Ch., *Epistemic logic*, (in:) L. Goble (ed.), *The Blackwell Guide to Philosophical Logic*, Oxford, Blackwell Publichers, 2001, pp. 183–202.

[190] Michael F. S., *What is the master argument of Diodorus Cronus?*, American Philosophical Quarterly, 13 (1976), pp. 229–235.

[191] Michalski K., *Logika i czas. Próba analizy Husserlowskiej teorii sensu* (*Logic and Time. An Approach to Husserl's Theory of Sense*), Warszawa, PIW, 1988.

[192] Mill J. S., *A System of Logic*, Honolulu, University Press of the Pacific, 2002.

[193] Montague R., *Pragmatics*, (in:) R. Klibansky (ed.), *Contemporary Philosophy: A Survey*, Florence, La Nuova Italia Editrice, 1968, pp. 102–122.

[194] Montague R., *Pragmatics and intensional logics*, Synthese, 22 (1970), pp. 68–94.

[195] Moore G. E., *Some Main Problems of Philosophy*, London, Allen and Unwin, 1953.

[196] Morris Ch. W., *The Foundations of the Theory of Signs*, Chicago, University of Chicago Press, 1938.

[197] Mostiepanienko A., *Wymiar przestrzeni a następstwo czasowe* (*Dimension of space and temporal succession*), M. Korzeniowska (trans. from Russian), (in:) *Przestrzeń, czas, ruch* (*Space, Time, Motion*), Warszawa, PWN, pp. 38–62.

[198] Mrugalski Z., *Mechanizmy zegarowe* (*Clock Mechanisms*), Warszawa, WNT, 1972.

[199] Mulligan K., P. Simons and B. Smith, *Thruth-makers*, Philosophy and Phenomenological Research, 3 (1984), pp. 287–321.

[200] Nasieniewski M., *Rekonstrukcja logiki stoickiej jako rachunku sekwentów* (*A reconstruction of stoic logic as a sequent calculus*), (in:) J. Perzanowski and A. Pietruszczak (eds.), *Logika i Filozofia Logiczna. 1996–1998*, Toruń, UMK, 2000, pp. 207–242.

[201] Neurath O., *Gesammelte philosophische und methodologishe Schriften*, vol. 2, Wien, Hölder-Pichler-Tempsky, 1981.

[202] Newton R. G., *What Makes Nature Tick?*, Cambridge, Harvard University Press, 1998.

[203] Nowikow I., *Rzeka czasu. Czarne dziury, białe dziury i podróże w czasie* (*River of Time. Black Holes, White Holes and Time Travels*), Warszawa, Prószyński i S-ka, 2001.

[204] Øhstrøm P., *A new reconstruction of the master argument of Diodorus Cronus*, International Logic Review, 1 (1980), pp. 60–65.

[205] Øhstrøm P. and P. F. V. Hasle, *Temporal Logic. From Ancient Ideas to Artificial Intelligence*, Dordrecht, Kluwer Academic Publishers, 1995.

[206] Ogden C. K. and I. A. Richards, *The Meaning of Meaning*, New York, Harvest Book, 1923.

[207] Ogrodnik B., *Ontologia czasu konkretnego* (*Ontology of a Specified Time*), Katowice, Wydawnictwo Uniwersytetu Śląskiego, 1995.

[208] Omyła M., *Zarys logiki niefregowskiej* (*Outline of Non-Fregean Logic*), Warszawa, PWN, 1986.

[209] Panikkar R., *Czas i historia w tradycji indyjskiej: Kala i karman*, K. Byrski (trans.), (in:) A. Zajączkowski (ed.), *Czas w kulturze*, Warszawa, PIW, 1988, pp. 111–138. Eng. version: *Time and history in the Indian tradition: Kala and Karman*, (in:) L. Gardet (ed.), *Cultures and Time*, Paris, The Unesco Press, 1976.

[210] Pawełczyńska A., *Czas i przestrzeń a formy poznania świata* (*Time and space vs. forms of cognitions of world*), (in:) S. Nowak (ed.), *Wizje człowieka i społeczeństwa w teoriach i badaniach naukowych* (*An Image of Human and Society in Scientific Theories and Researches*), Warszawa, PWN, 1984, pp. 279–305.

[211] Pàttaro G., *Pojmowanie czasu w chrześcijaństwie*, B. Chwedeńczuk (trans.), (in:) A. Zajączkowski (ed.), *Czas w kulturze*, Warszawa, PIW, 1988,

pp. 291–329. Eng. version: *The christian conception of time*, (in:) L. Gardet (ed.), *Cultures and Time*, Paris, The Unesco Press, 1976.

[212] Peirce Ch. S., *How to make our ideas clear?*, Popular Science Monthly, 12 (1878), pp. 286–302.

[213] Perzanowski J. and A. Pietruszczak, *Aim, the Scope and Editorial Policy of the Journal*, Logic and Logical Philosophy, 1 (1993), pp. 3–6.

[214] Perzanowski J., *Logiki modalne a filozofia* (*Modal logic and philosophy*), (in:) J. Perzanowski (ed.), *Jak filozofować? Studia z metodologii filozofii*, Warszawa, PWN, 1989, pp. 262–346.

[215] Perzanowski J., *Ontological arguments II: Cartesian and Leibnizian*, (in:) H. Burkhardt, B. Smith (eds.), *Handbook of Methaphisycs and Ontology*, t. II, Munich/Philadelphia/Vienna, Philosophia Verlag, 1991, s. 625–633.

[216] Perzanowski J., *Reasons and causes*, (in:) J. Faye, U. Scheffler and M. Urchs (eds.), *Logic and Causal Reasoning*, Berlin, Akademie Verlag, 1994, pp. 169–191.

[217] Pietruszczak A., *Co to jest mereologia* (*What is mereology*), (in:) C. Gorzka and R. Jadczak (eds.), *Studia z filozofii i logiki*, Toruń, Wydawnictwo UMK, 1997, pp. 115–127.

[218] Pietruszczak A., *Metamereologia* (*Metamereology*), Toruń, Wydawnictwo UMK, 2000.

[219] Piłat R., *Przeżycie chwili obecnej* (*Experience of the present moment*), Studia z kognitywistyki i filozofii umysłu, 1 (2003), pp. 61–84.

[220] Placek T., *Time and passing*, (in:) M. Heller and M. Mączka (eds.), *Czas…* (*Time…*), Kraków, Biblos, 2001, pp. 84–102.

[221] Placek T., *Is Nature Deterministic?*, Kraków, Jagiellonian University Press, 2000.

[222] Plato, *Sophist*, N. White (trans.), (in:) J. M. Cooper (ed.), *Plato: Complete Works*, Indianapolis, Hackett Publishing, 1997, pp. 235–293.

[223] Podgórski K., *Determinizm a mechanika klasyczna* (*Determinism and classical mechanics*), Studia Filozoficzne, 5 (1986), pp. 55–60.

[224] Popper K., *A World of Propensities: The New Views of Causality*, Bristol, Thoemmes Antiquarian Books Ltd., 1990.

[225] Popper K., *The Open Universe. An Argument for Indeterminism*, London, Routledge, 1988.

[226] Price H., *Time's Arrow and Archimedes' Point*, New York/Oxford, Oxford University Press, 1997.

[227] Priest G., *An Introduction to Non-Classical Logic*, Cambridge, Cambridge University Press, 2001.

[228] Prior A. N., *Diodorean modalities*, Philosophical Quarterly, 5 (1955), pp. 205–213.

[229] Prior A. N., *Past, Present and Future*, Oxford, Clarendon Press, 1967.

[230] Prior A. N., *Time and Modality*, Oxford, Clarendon Press, 1957.

[231] Prior A. N., *Tense-logic and the continuity of time*, Studia Logica, 13 (1962), p. 133–148.

[232] Putnam H., *The Many Faces of Realism: The Paul Carus Lectures*, La Salle, Open Court, 1987.

[233] Quine W. V. O., *Notes on the theory of reference*, (in:) *From a Logical Point of View*, Cambridge/London, Harvard University Press, 2003, pp. 130–138.

[234] Quine W. V. O., *Two dogmas of empiricism*, (in:) *From a Logical Point of View*, Cambridge/London, Harvard University Press, 2003, pp. 20–46.

[235] Quine W. V. O., *Word and Object*, Cambridge, The MIT Press, 1960.

[236] Reale G., *A History of Ancient Philosophy*, vol. 1 and 2, J. R. Catan (trans.), New York, State University of New York Press, 1987, 1990.

[237] Reichenbach H., *The Direction of Time*, Berkeley/Los Angeles, University of California Press, 1956.

[238] Reichenbach H., *Elements of Symbolic Logic*, New York, Macmillan, 1947.

[239] Reichenbach H., *Kausalstruktur der Welt und der Unterschied von Vergangenheit und Zukunft*, Sitzungsberichte der Bayerischen Akademie der Wissenschaft, 1925, pp. 133–175.

[240] Rescher N., *On the logic of chronological propositions*, Mind, 77 (1966), pp. 75–96.

[241] Rescher N. and A. Urquhart, *Temporal Logic*, Vienna, Springer-Verlag, 1971.

[242] Rescher N., *Truth and necessity in temporal perspective* (in:) R. Gale (ed.), *The Philosophy of Time*, New Jersey, MacMillan, 1968, pp. 183–220.

[243] Rescher N., *A version of the Master Argument of Diodorus*, Journal of Philosophy, 43 (1966), pp. 438–445.

[244] Rorty R., *Philosophy and the Mirror of Nature*, Princeton, Princeton University Press, 1979.

[245] Russell B., *On denoting*, (in:) *Logic and Knowledge*, London, Allen and Unwin, 1956, pp. 41–56.

[246] Russell B., *Descriptions*, (in:) *Introduction to Mathematical Philosophy*, London, Allen and Unwin, 1919, pp. 167–180.

[247] Russell B., *Human Knowledge. Its Scope and Limits*, London, Routledge, 1994.

[248] Russell B., *An Inquiry into Meaning and Truth*, London, Routledge, 1995.

[249] Russell B., *The philosophy of logical atomism*, (in:) D. F. Pears (ed.), *The Philosophy of Logical Atomism*, Lasalle, Open Court, 1985, pp. 35–155.

[250] Russell B., *The Problems of Philosophy*, London, Wiliam and Norgate, 1912.

[251] Salmon W. C., *Scientific Explanation and the Causal Structure of the World*, Princeton, Princeton University Press, 1984.

[252] Saarinen E., *Backwards-looking operators in tense logic and in natural language*, (in:) J. Hintikka, I. Niiniluoto and E. Saarinen (eds.), *Essays on Mathematical and Philosophical Logic*, Dordrecht, Reidel Publishing Company, 1978, pp. 341–368.

[253] Schlick M., *Die Kausalität in der gegenwärtigen Physik*, Die Naturwissenschaften, 19 (1931), pp. 145–162.

[254] Sedley D., *Hellenistic physics and metaphysics*, (in:) K. Algra, J. Barnes, J. Mansfeld and M. Schofield (eds.), *The Cambridge History of Philosophy*, Cambridge, Cambridge University Press, 1999, pp. 355–411.

[255] Sellars W., *Science and Methaphysics*, London, Routledge and Kegan, 1968.

[256] Schrödinger E., *Irreversibility*, Proceedings of the Royal Irish Academy, 53 (1950–1951), pp. 189–195.

[257] Scott D., *Advice on modal logic*, (in:) K. Lambert (ed.), *Philosophical Problems in Logic*, Dordrecht, Reidel, 1970, pp. 143–173.

[258] Simons P., *Moments as truth-makers*, (in:) W. Leinfellner (ed.), *Language and Ontology*, Proceedings of the Sixth International Wittgenstein Symposium 1981, Vienna, Hölder-Pichler-Tempsky, 1982, pp. 159–161.

[259] Smart J. J. C., *Causal theories of time*, (in:) *Essays Metaphysical and Moral. Selected Philosophical Papers*, Oxford/New York, Basil Blackwell, 1994, pp. 51–60.

[260] Smart J. J. C., *The reality of the future*, (in:) *Essays Metaphysical and Moral. Selected Philosophical Papers*, Oxford/New York, Basil Blackwell, 1994, pp. 91–99.

[261] Snihur S., *Kilka uwag w sprawie potocznego pojmowania czasu* (*Some remarks concerning a common sense understanding of time*), Studia Filozoficzne, 9 (1986), pp. 39–46.

[262] Spinoza B., E. Curley (trans.) *Ethics*, (in:) E. Curley (ed.), *The Collected Works of Spinoza*, vol. 1, Princeton, Princeton University Press, 1985, pp. 408–620.

[263] Stalnaker R., *Pragmatics*, Synthese, 22 (1970), pp. 272–289.

[264] Such J., *Czas i przestrzeń* (*Time and space*), (in:) Z. Cackowski, J. Kmita, K. Szaniawski and P. J. Smoczyński (eds.), *Filozofia a nauka. Zarys encyklopedyczny* (*Time and Science. Encyclopaedic Outline*), Wrocław, PAN, 1987, pp. 61–75.

[265] Such. J, *Prawo naukowe (i prawidłowość) (Scientific laws (and regularities))*, (in:) Z. Cackowski, J. Kmita, K. Szaniawski and P. J. Smoczyński (eds.), *Filozofia a nauka. Zarys encyklopedyczny (Time and Science. Encyclopaedic Outline)*, Wrocław, PAN, 1987, pp. 519–532.

[266] Surowik D., *Intuicjonistyczna logika i indeterminizm (Intuitionistic logic and indeterminism)*, PhD dissertation promoted by K. Trzęsicki, Białystok, 2001.

[267] Szubka T., *Antyrealizm semantyczny. Studium analityczne (Semantic Anti-realism. Analytical Study)*, Lublin, RW KUL, 2001.

[268] Szumilewicz I., *O kierunku upływu czasu (On Direction of Passing Time)*, Warszawa, PWN, 1964.

[269] Święcicka K., *Husserl*, Warszawa, Wiedza Powszechna, 1993.

[270] Świrydowicz K., *Podstawy logiki modalnej (The Basics of Modal Logic)*, Poznań, Wydawnictwo Naukowe UAM, 2004

[271] Tarski A., *The concept of truth in formalized languages*, J.H. Woodger (trans.), (in:) J. Corcoran (ed.), *Logic, Semantics, Metamathematics*, Indianapolis, Hackett, 1983, pp. 152–278.

[272] Tarski A., *The semantic conception of truth and the foundations of semantics*, Philosophy and Phenomenological Research, 4 (1944), pp. 341–376.

[273] Tarski A., *Truth and proof*, Scientific American, 220, 6 (1969), pp. 63–77.

[274] Tarski A., *Prawda i fałsz (Truth and falsity)*, Studia filozoficzne, 2 (1984), pp. 21–22.

[275] Taylor E. F. and J. A. Wheeler, *Spacetime Physics*, New York, W. H. Freeman and Co., 1992.

[276] Tempczyk M., *Fizyka a świat realny. Elementy filozofii fizyki (Physics vs. Reality. Elements of Philosophy of Physics)*, Warszawa, PWN, 1986.

[277] *The Christian Standard Bible*, Holman Bible Publishers, 2017.

[278] Thomas A., *The Summa contra Gentiles*, English Dominican Fathers (trans.), (in:) *The Summa contra Gentiles of Saint Thomas Aquinas*, 1st and 2nd book, London, Burns, Oates & Washbourne Ltd., 1924, 1923.

[279] Thomas A., *Treatise on God*, J. F. Anderson (trans.), Englewood Cliffs, Prentice-Hall, 1963.

[280] Tkaczyk M., *Logika czasu empirycznego (Logic of Empirical Time)*, Lublin, Wydawnictwo KUL, 2009.

[281] Tokarz M., *Elementy formalnej teorii składni (Elements of Formal Theory of Syntax)*, Warszawa, Biblioteka Myśli Semiotycznej, 1994.

[282] Tokarz M., *Elementy pragmatyki logicznej (Elements of Logical Pragmatics)*, Warszawa, PWN, 1993.

[283] Trzęsicki K., *Intuitionism and indeterminism (tense-logical considerations)*, (in:) J. Woleński (ed.), *Philosophical Logic in Poland*, Dordrecht, Kluwer Academic Publishers, 1994, pp. 271–296.

[284] Trzęsicki K., *Is discreteness of time necessary for Diodorean Master Argument?*, Bulletin of the Section of Logic, 16 (1987), pp. 125–135.

[285] Trzęsicki K., *Tense-logical approach to the problem of determinism and causality*, (in:) *Tense-logical Approach to the Problem of Determinism* (typescript), Białystok, 1989, pp. 5–72.

[286] Vendler Z., *Verbs and times*, Philosophical Review, 66 (1957), pp. 143–160.

[287] Venema Y., *Temporal logic*, (in:) L. Goble (ed.), *The Blackwell Guide to Philosophical Logic*, Oxford, Blackwell Publishers, 2001, pp. 203–223.

[288] Weyl H., *Philosophie der Mathematik und Naturwissenschaft*, M'nchen, R. Oldenburg, 1926.

[289] Whorf B. L., *Language, Thought and Reality*, Boston, MIT Press, 1956.

[290] Williamson T., *Knowledge and Its Limits*, Oxford, Oxford University Press, 2000.

[291] Wittgenstein L., *Tractatus Logico-Philosophicus*, C. K. Ogden (trans.), London, Routledge & Kegan Paul, 1992.

[292] Wittgenstein L., *Philosophical Investigations*, G. E. M. Anscombe (trans.), G. E. M. Anscombe and R. Rhees (eds.), Oxford, Blackwell, 1953.

[293] West M. L., *Hesiod Works and Days*, Oxford, Clarendon Press, 1978.

[294] Woleński J., *Epistemologia. Zarys historyczny i problemy metateoretyczne* (*Epistemology, Historical Outline and Metatheoretic Problems*), vol. 1, Kraków, Aureus, 2000.

[295] Woleński J., *Metamatematyka a epistemologia* (*Metamathematics and Epistemology*), Warszawa, PWN, 1993.

[296] Wolniewicz B., *Ontologia sytuacji* (*Ontology of Situations*), Warszawa, PWN, 1985.

[297] Wójcicki R., *Metodologia formalna nauk empirycznych* (*Formal Methodology of Science*), Wrocław, Ossolineum, 1974.

[298] Wright von G. H., *And next*, Acta Philosophica Fennica, 18 (1965), pp. 293–304.

[299] Wright von G. H., *Norm and Action*, New York/London, Routledge and Kegan Paul, 1963.

[300] Wright von G. H., *On the logic and epistemology of causal relation*, (in:) E. Sosa and M. Tooley (eds.), *Causation*, New York, Oxford University Press, 1993, pp. 105–124.

[301] Zajączkowski A., *Czas Afryki czarnej* (*Time of black Africa*), (in:) A. Zajączkowski (ed.), *Czas w kulturze* (*Time in Culture*), Warszawa, PIW, 1988, pp. 15–46.

[302] Zajączkowski A., *Wstęp* (*Introduction*), (in:) A. Zajączkowski (ed.) *Czas w kulturze* (*Time in Culture*), Warszawa, PIW, 1988, pp. 5–11.

[303] Zajdler L., *Dzieje zegara* (*History of Clock*), Warszawa, Wiedza Powszechna, 1980.

[304] Zeller E., *Die Philosophie der Griechen in ihrer geschichtlichen Entwicklung*, vol. 2, Leipzig, Fues's Verlag, 1875.

[305] Zeller E., *Über den κυριευων des megarikers Diodorus*, (in:) *Sitzungsberichte der Königlichen Akademie der Wissenschaften zu Berlin*, Berlin, Reichsdruckerei, 1882, pp. 151–159.

[306] Żegleń U., *Modalność w logice i filozofii. Podstawy ontyczne* (*Modalities in Logic and Philosophy. Ontological Fundamentals*), Warszawa, Polskie Towarzystwo Semiotyczne, 1990.

Index

T

DIA-LOGOS
Schriften zu Philosophie und Sozialwissenschaften
Studies in Philosophy and Social Sciences

Herausgegeben von
Edited by
Tadeusz Buksiński & Piotr W. Juchacz

"*Dia-Logos. Studies in Philosophy and Social Sciences* is a peer-reviewed book series publishing valuable monographs and edited volumes on various aspects of philosophy and social sciences. The series is intended to be an interdisciplinary forum of deliberation according to our firm belief that challenges of the contemporary world require common and multilevel research. The *Dia-Logos* series does not represent a single ideology or school of thought, but it is open to different trends and various styles of reflection, trying to understand better the contemporary world. We invite the submission of manuscripts of monographs and edited volumes from academic philosophers and social scientists."

Bd. / Vol. 1 Piotr W. Juchacz / Roman Kozłowski (Hrsg.): Freiheit und Verantwortung. Moral, Recht und Politik. 2002.

Bd. / Vol. 2 Norbert Leśniewski / Ewa Nowak-Juchacz (Hrsg.): Die Zeit Heideggers. 2002.

Bd. / Vol. 3 Marek Kwiek (ed.): The University, Globalization, Central Europe. 2003.

Bd. / Vol. 4 Ewa Czerwińska-Schupp (Hrsg.): Philosophie an der Schwelle des 21. Jahrhunderts. Geschichte der Philosophie, Philosophische Anthropologie, Ethik, Wissenschaftstheorie, Politische Philosophie. 2003.

Bd. / Vol. 5 Danuta Sobczyńska / Pawel Zeidler / Ewa Zielonacka-Lis (eds.): Chemistry in the Philosophical Melting Pot. 2004.

Bd. / Vol. 6 Marek Kwiek: Intellectuals, Power, and Knowledge. Studies in the Philosophy of Culture and Education. 2004.

Bd. / Vol. 7 Marek Kwiek: The University and the State. A Study into Global Transformations. 2006.

Bd. / Vol. 8 Andrzej Przylebski (Hrsg.): Das Erbe Gadamers. 2006.

Bd. / Vol. 9 Ewa Czerwińska-Schupp (ed.): Values and Norms in the Age of Globalization. 2007.

Bd. / Vol. 10 Tadeusz Buksiński (ed.): Democracy in Western and Post-Communist Countries. Twenty Years after the Fall of Communism. 2009.

Bd. / Vol. 11 Marek Zirk-Sadowski / Mariusz Golecki / Bartosz Wojciechowski (eds.): Multicentrism as an Emerging Paradigm in Legal Theory. 2009.

Bd. / Vol. 12 Bartosz Wojciechowski: Philosophical Approach to the Interculturality of Criminal Law. 2010.

Bd. / Vol. 13 Marek Nowak / Michał Nowosielski (eds.): (Post)transformational Migration. Inequalities, Welfare State, and Horizontal Mobility. 2011.

Bd. / Vol. 14 Tadeusz Buksiński (ed.): Religions in the Public Spheres. 2011.

Bd. / Vol. 15 Bartosz Wojciechowski / Piotr W. Juchacz / Karolina M. Cern (eds.): Legal Rules, Moral Norms and Democratic Principles. 2013.

Bd. / Vol. 16 Ewa Nowak / Dawn E. Schrader / Boris Zizek (eds.): Educating Competencies for Democracy. 2013.

Bd. / Vol. 17 Tadeusz Buksiński (ed.): Identities and Modernizations. 2013.

Bd. / Vol. 18 Karolina M. Cern: The Counterfactual Yardstick. Normativity, Self-Constitutionalisation and the Public Sphere. 2014.

Bd. / Vol. 19 Vito Breda / Lidia Rodak (eds.). Diverse Narratives of Legal Objectivity. An Interdisciplinary Perspective. 2016.

Bd. / Vol. 20 Marek Woszczek: Platonic Wholes and Quantum Ontology. Translated by Katarzyna Kretkowska. 2015.

Bd./Vol. 21 Roman Hauser / Marek Zirk-Sadowski / Bartosz Wojciechowski (eds.): The Common European Constitutional Culture. Its Sources, Limits and Identity. 2016.

Bd./Vol. 22 Aigerim Raimzhanova: Hard, Soft, and Smart Power – Education as a Power Resource. 2017.

Bd./Vol. 23 Roberto Franzini Tibaldeo / Graziano Lingua (eds.): Philosophy and Community Practices. 2018.

Bd./Vol. 24 Marta Soniewicka: After God – The Normative Power of the Will from the Nietzschean Perspective. 2017.

Bd./Vol. 25 Tomasz Jarmużek: On the Sea Battle Tomorrow That May Not Happen. A Logical and Philosophical Analysis of the Master Argument. 2018.

www.peterlang.com

www.ingramcontent.com/pod-product-compliance
Ingram Content Group UK Ltd.
Pitfield, Milton Keynes, MK11 3LW, UK
UKHW041825031026
15587UKWH00002B/559

9 783631 745892